D1217536

DFG Deutsche Forschungsgemeinschaft

The MAK-Collection
for Occupational Health and Safety

Part III: Air Monitoring Methods

Volume 9

edited by Harun Parlar
Working Group Analytical Chemistry

Commission for the Investigation of Health
Hazards of Chemical Compounds in the Work Area
(Chairman: Helmut Greim)

WILEY-
VCH

WILEY-VCH Verlag GmbH & Co. KGaA

Prof. Dr. med. Helmut Greim
Senatskommission
zur Prüfung gesundheitsschädlicher Arbeitsstoffe
der Deutschen Forschungsgemeinschaft
Technische Universität München
Hohenbachernstr. 15–17
85354 Freising-Weihenstephan
Germany

Prof. Dr. Dr. Harun Parlar
Technische Universität München
Wissenschaftszentrum Weihenstephan
für Ernährung, Landnutzung und Umwelt
Lehrstuhl für Chemisch-Technische Analyse u.
Chemische Lebensmitteltechnologie
Weihenstephaner Steig 23
85350 Freising-Weihenstephan
Germany

Translator: Dr. J. Cito Habicht

Volumes 1–8 were published under the title "Analyses of Hazardous Substances in Air" (ISSN 0946-7610)

Library of Congress Card No.: applied for
A catalogue record for this book is available from the British Library.
Die Deutsche Bibliothek – CIP Cataloguing-in-Publication-Data
A catalogue record for this publication is available from Die Deutsche Bibliothek

ISSN: 1860-4986
ISBN-13: 978-3-527-31138-8
ISBN-10: 3-527-31138-6

© 2005 WILEY-VCH Verlag GmbH & Co. KGaA, Weinheim

Printed on acid-free paper.

Composition: ProSatz Unger, Weinheim
Printing: Strauss Offsetdruck, Mörlenbach
Bookbinding: Litges & Dopf Buchbinderei GmbH, Heppenheim

Printed in the Federal Republic of Germany.

Preface

This is the first volume of the new international series "The MAK-Collection for Occupational Health and Safety" which offers as of this volume all the comprehensive toxicological documentations as well as validated analytical methods issued by the Commission for the Investigation of Health Hazards of Chemical Compounds in the Work Area of the Deutsche Forschungsgemeinschaft (DFG). The Commission and the publisher Wiley-VCH hope that the new, concise title and a regular annual publication schedule will even better meet the worldwide demand for reliable, unbiased information about hazardous compounds in the work area. The new series, featuring an attractive cover design, is published in four parts, this volume belonging to Part III: Air Monitoring Methods. The volume is published concurrent with the German edition "Analytische Methoden zur Prüfung gesundheitsschädlicher Arbeitsstoffe". Previously, the volumes were published in the series "Analyses of Hazardous Substances in Air". Since the beginning of the nineties, when the first volume appeared, eight volumes have been published. On this occasion my special thanks are due to the Deutsche Forschungsgemeinschaft, which has lent its support to the German and English collection of methods for more than three decades.

The current volume is being published by the DFG Working Subgroup "Analysis of Hazardous Substances in Air of Work Areas" in collaboration with the Analytical Working Group of the Expert Committee "Chemistry" of the Employment Accident Insurance Fund of the Chemical Industry (Berufsgenossenschaft der chemischen Industrie). The Commission hopes that by publishing both sets of methods – analysis in air of hazardous substances with MAK values, so-called DFG methods and analysis in air of carcinogenic substances, so-called BGI 505 procedures – in one English volume, the growing repertoire of methods will be put to effective use, e.g. within the European Union in the efforts to protect health at the workplace.

The concept of this volume is similar to that of the preceding volumes. It comprises thirteen new analytical methods, which enable the monitoring of concentrations of different hazardous substances in the air of the workplace area, as well as a chapter on the sampling and determining of aerosols and their chemical components. This chapter provides a broad overview of the sampling procedures for dusts and the subsequent quantitative determination of different dust fractions and metal-containing components of dust.

It is gratifying for me, as chairman of the Deutsche Forschungsgemeinschaft's Commission for the Investigation of Health Hazards of Chemical Compounds in the Work Area, to ascertain the progress that we have made towards our goal concerning the protection of workers against risks from chemical agents in work areas. For this reason I would like to express my gratitude to Prof. Dr. A. Kettrup, who has successfully headed the Working Subgroup "Analyses of Hazardous Substances in Air of Work Area" for many years, and Prof. Dr. H. Parlar, who assumed the chairmanship at the end of 2003, for their successful work.

The MAK-Collection Part III: Air Monitoring Methods, Vol. 9. DFG, Deutsche Forschungsgemeinschaft
Copyright © 2005 WILEY-VCH Verlag GmbH & Co. KGaA, Weinheim
ISBN: 3-527-31138-6

Further, I extend my thanks to the authors and examiners of these methods for their contributions, as well the translator of this volume, Dr. J. Cito Habicht. Particular thanks go to two members of the Secretariat of the Commission – Dr. M. R. Lahaniatis and Dr. R. Schwabe – for the successful continuation of their work and their personal engagement.

H. Greim
Chairman of the Commission for the
Investigation of Health Hazards of Chemical
Compounds in the Work Area

Foreword

We are pleased to present a new volume of "The MAK-Collection for Occupational Health and Safety", Part III: Air Monitoring Methods, with thirteen new analytical procedures, applicable for routine determination of hazardous substances in the air of working places. The development and verification of these kinds of methods is a long-term objective of the Working Subgroup "Analyses of Hazardous Substances in Air of Work Areas" in keeping with the goals of the Working Group "Analytical Chemistry" of the Commission for the Investigation of Health Hazards of Chemical Compounds in the Work Area of the Deutsche Forschungsgemeinschaft (DFG).

The analytical procedures presented in this volume were developed by the experts of the DFG Working Subgroup "Analyses of Hazardous Substances in Air of Work Area" and the Analytical Working Group of the Board of Experts "Chemistry" of the Berufs-genossenschaften (Employment Accident Insurance Fund of the Chemical Industry), and have been tested and verified by at least one examiner for their reliability and reproducibility.

This volume contains, among other methods, ones for basic chemicals such as ammonia, sulfuric acid, and hydrogen fluoride and fluorides which play an important role in working areas. With the developed methods these substances can be easily determined in concentration, even significantly lower than the threshold levels. Trichloroethylene and tetrachloroethylene are critical solvents in the chemical industry and in dry cleaning factories. In order to measure their concentration a rapid and reliable method was recommended, which is characterized by its reproducibility. In this volume, the analytical determination methods for the substances, 2-butenal, halogenated anaesthetic gases (halothane, enflurane, isoflurane) and dicyclopentadiene, all with high vapour pressure, are included. Quantification of these substances in air is very rapid and allows short-time period checking in working areas and in operating rooms. Additionally, the analytical procedures for the determination of 2-butanone oxime, nitrotoluene (2,4-dinitrotoluene, 2,6-dinitrotoluene and 2,4,6-trinitrotoluene) and triglycidyl isocyanurate in air are included in this volume. Analytical processes for these compounds, also in regard to their threshold limit values, are reliable and suitable for correct determination of these substances in air.

This volume contains, in addition to the thirteen analytical methods, a chapter on the sampling and determining of aerosols and their chemical components, as well as a list of the Members, Guests and ad hoc Experts of the DFG Working Subgroup "Analyses of Hazardous Substances in Air of Work Area", a list of the Members of the Analytical Working Group of the Board of Experts "Chemistry", and an index of the CAS numbers of the investigated substances.

The MAK-Collection Part III: Air Monitoring Methods, Vol. 9. DFG, Deutsche Forschungsgemeinschaft
Copyright © 2005 WILEY-VCH Verlag GmbH & Co. KGaA, Weinheim
ISBN: 3-527-31138-6

We would like to thank all the experts and the reviewers for their efforts. We thank the DFG for financial and organisational support for our activities.

J. Angerer
Chairman of the Working Group
"Analytical Chemistry" of the
Commission for the Investigation
of Health Hazards of Chemical
Compounds in the Work Area

H. Parlar
Chairman of the Working Subgroup
"Analyses of Hazardous Substances in Air
of Work Areas"

Contents

The MAK-Collection Part III: Air Monitoring Methods, Vol. 9. DFG, Deutsche Forschungsgemeinschaft
Copyright © 2005 WILEY-VCH Verlag GmbH & Co. KGaA, Weinheim
ISBN: 3-527-31138-6

Working Group "Analytical Chemistry" of the Commission for the Investigation of Health Hazards of Chemical Compounds in the Work Area of the Deutsche Forschungsgemeinschaft

Organization

The Working Group "Analytical Chemistry" was established in 1969. Under the chairmanship of Prof. Dr. J. Angerer at the present it includes two Working Subgroups:

"Air Analyses"
(Leader: Prof. Dr. H. Parlar)

"Analyses of Hazardous Substances in Biological Materials"
(Leaders: Prof. Dr. J. Angerer and Chem.-Ing. K. H. Schaller).

The participants, who have been invited to collaborate on a Working Subgroup by the leaders, are experts in the field of technical and medical protection against chemical hazards at the workplace.
A list of members and guests of "Analyses of Hazardous Substances in Air" is given at the end of this volume.

Objectives and operational procedure

The two analytical subgroups are charged with the task of preparing methods for the determination of hazardous industrial materials in the air of the workplace or to determine these hazardous materials or their metabolic products in biological specimens from the persons working there. Within the framework of the existing laws and regulations, these analytical methods are useful for ambient monitoring at the workplace and biological monitoring of the exposed persons.
In addition to working out the analytical procedure, these subgroups are concerned with the problems of the preanalytical phase (specimen collection, storage, transport), the statistical quality control, as well as the interpretation of the results.

Development, examination, release, and quality of the analytical methods

In its selection of suitable analytical methods, the Working Group is guided mainly by the relevant scientific literature and the expertise of the members and guests of the Working Subgroup. If appropriate analytical methods are not available they are worked out within the Working Group. The leader designates an author, who assumes the task of developing and formulating a method proposal. The proposal is examined experimentally by at least one other member of the project, who then submits a written report of the results of the examination. As a matter of principle the examination must encom-

The MAK-Collection Part III: Air Monitoring Methods, Vol. 9. DFG, Deutsche Forschungsgemeinschaft
Copyright © 2005 WILEY-VCH Verlag GmbH & Co. KGaA, Weinheim
ISBN: 3-527-31138-6

pass all phases of the proposed analytical procedure. The examined method is then laid before the members of the subgroups for consideration. After hearing the judgement of the author and the examiner they can approve the method. The method can then be released for publication after a final meeting of the leader of the Working Group "Analytical Chemistry" with the subgroup leaders, authors, and examiners of the method.

Under special circumstances an examined method can released for publication by the leader of the Working Group after consultation with the subgroup leaders.

Only methods for which criteria of analytical reliability can be explicitly assigned are released for publication. The values for inaccuracy, imprecision, detection limits, sensitivity, and specificity must fulfil the requirements of statistical quality control as well as the specific standards set by occupational health. The above procedure it meant to guarantee that only reliably functioning methods are published, which are not only reproducible within the framework of the given reliability criteria in different laboratories, but also can be monitored over the course of time.

In the selection and development of a method for determining a particular substance the Working Group has given the analytical reliability of the method precedence over aspects of simplicity and economy.

Publications of the working group

Methods released by the Working Group are published in the Federal Republic of the Germany, by the Deutsche Forschungsgemeinschaft as a loose-leaf collection entitled "Analytische Methoden zur Prüfung gesundheitsschädlicher Arbeitsstoffe" (Wiley-VCH Verlag, Weinheim, FRG).

The collection at present consists of two volumes:

Volume I   "Luftanalysen"
Volume II   "Analysen in biologischem Material".

These methods are also to be published in an English edition. Volume 1 to 9 of "Analyses of Hazardous Substances in Biological Materials" have already been published. The work at hand represents the ninth English issue of "Analyses of Hazardous Substances in Air".

Withdrawal of methods

An analytical method that is made obsolete by new developments or discoveries in the fields of instrumental analysis or occupational health and toxicology can be replaced by a more efficient method. After consultation with the membership of the relevant project and with the consent of the leader of the Working Group, the subgroup leader is empowered to withdraw the old method.

Analytical Working Group of the Board of Experts "Chemistry" of the Berufsgenossenschaften (Employment Accidents Insurance Institutions of Germany)

Organization and aims

In 1978 the Board of Experts "Chemistry" created a Working Group named "Analysis" with the aim of evaluating the practicability of existing procedures for measuring carcinogenic substances of practical and technical importance. It was discovered that practicable procedures existed for only a few carcinogenic substances.

New procedures were therefore developed on the basis of existing laws and regulations and published as procedures validated by the Berufsgenossenschaften by Carl Heymanns Verlag KG, Cologne, with the number BGI 505.- (ZH1/120.- formerly).

In addition to the analytical determinations, the methods developed also include information on sampling and the shelf life of the samples collected.

The members and guests of the Analytical Working Group are experts in the fields of the measurement and evaluation of hazardous substances and chemical analysis and are drawn from industry, regional authorities and research institutes.

A list of members of the Working Group "Analysis" is given at the end of this volume.

Development, testing, validation and quality of the analytical procedures

After a substance has been classed as carcinogenic in the national regulations, the chairman nominates as author a member of the working group or a guest expert for the development of a measuring procedure.

Standard texts drawn up by the working group serve as an aid in the layout and formulation of the suggested method.

In several readings the members of the working group evaluate and discuss the author's suggestion. Possible amendments or changes, i.e. explanations with the aid of existing data or if necessary from additional laboratory tests, finally lead to the description of a measuring procedure which is accepted from all working group members. The method is then approved for publication by the chairman. Procedures are only published by the Berufsgenossenschaften as approved methods when they fulfil the recognised requirements of statistical quality control with regard to quantification limit, accuracy, selectivity and sensitivity. In addition, the trueness of the measuring procedure must be guaranteed in comparative studies or using test gas. The overall uncertainty resulting from all the systematic and random errors that occur during measurement must not exceed 30%. The concentration range for the particular hazardous substance to be measured using the procedure must be between a tenth, if not realizable a fifth, and three times the limit value.

Up-dating the analytical existing methods

At regular intervals existing methods (at present 72 methods) are revised to take into account improvements in measuring procedures and the current limit values. An overview is given on the web pages of the Berufsgenossenschaft der chemischen Industrie (http://www.bgchemie.de).

Preliminary Remarks

The MAK-Collection Part III: Air Monitoring Methods, Vol. 9. DFG, Deutsche Forschungsgemeinschaft
Copyright © 2005 WILEY-VCH Verlag GmbH & Co. KGaA, Weinheim
ISBN: 3-527-31138-6

Sampling and determining aerosols and their chemical components

Contents

The MAK-Collection Part III: Air Monitoring Methods, Vol. 9. DFG, Deutsche Forschungsgemeinschaft
Copyright © 2005 WILEY-VCH Verlag GmbH & Co. KGaA, Weinheim
ISBN: 3-527-31138-6

1 Introduction

Aerosols are dispersions of solid or liquid particles in air. They include dusts, smokes and mists. The following chapter provides an overview of the sampling procedures for dusts and the subsequent quantitative determination of different dust fractions and metal-containing components of dust. This includes notes on how to conduct and evaluate dust measurements, select suitable sampling devices and sampling media and determine the inhalable and respirable fractions of dusts. Quantitative determination of chemical components of dust by means of suitable digestion and analysis methods will be discussed.

Workers and employees are exposed to dusts in many working environments. There are many different causes for their release. Some materials occur as powders or granulates in the first place. Handling by workers in most cases results in workplace exposure. However, dusts and smokes are also released during mechanical and thermal processing of surfaces. Similarly, motor vehicle exhausts and other combustion fumes also contain particles. The spraying of liquids is associated with the formation of droplets, which to some extent contain particles, as is the case e. g. when spraying varnish or lacquer. All of these particulate components occurring in workplace air are summarised under the term "dusts". The examples given in Table 1 illustrate that there are industries in which marked exposure to dust occurs. On the other hand, however, there are certain occupational activities that are associated with considerable workplace exposure to dust, irrespective of industry.

Airborne dusts are categorised as follows on the basis of particle size:

– ultra-fine fraction,
– respirable fraction (referred to in Germany as the "A" fraction, formerly "Feinstaub" (fine dust)),
– thoracic fraction,
– inhalable fraction (referred to in Germany as the "E" fraction, formerly "Gesamtstaub" (total dust)),
– coarse-disperse fraction.

Of relevance to the evaluation of exposure in workplaces at present are the respirable, thoracic and inhalable dust fractions [2]. In this context, the inhalable dust fraction also

Table 1. Examples of industries and occupations with exposure to dust [1].

Industries in which exposure to dust is mainly due to the handling of dusts or dust-generating materials	Industries in which exposure to dust is associated with the processing of materials
Building industry	Wood and plastics industries and skilled trades
Mining industry	
Quarry, sand, gravel, lime and gypsum mining industries	Textile industry
	Paper industry
Glass and ceramics industries	Welding
Foundry industry	Grinding and milling
	Mechanical processing
	Demolition work

encompasses the thoracic and respirable fractions. The thoracic dust fraction in turn includes the respirable fraction.

Figure 1 shows the conventions for the dust fractions according to EN 481 [2]. In addition to those dust fractions, the deposition characteristics according to the Johannesburg Convention [3] are also shown. Sampling devices that are in compliance with the latter convention also meet the EN 481 requirements for the respirable dust fraction.

Coarse-disperse and ultra-fine particles are not defined by EN 481, which means that no deposition characteristics exist for them. Ultra-fine particles are characterised by a diffusion equivalent diameter <100 nm [4]. This also includes aggregates and agglomerates. More recently, an increasing number of studies have been conducted to charac-

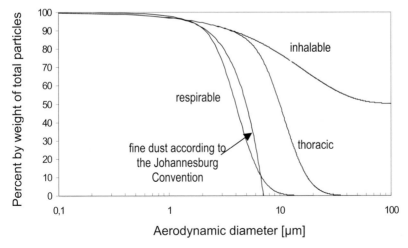

Fig. 1. Inhalable, thoracic and respirable conventions, in percent by weight of total airborne particles according to EN 481, and fine dust fraction according to the Johannesburg Convention.

terise exposure to ultra-fine dusts. The term "coarse-disperse particles" refers to the non-inhalable portions of airborne dust in the breathing zone, their aerodynamic diameter being greater than 100 μm. Exactly what particles are still in the inhalable range depends on the working conditions [5].

The limit value in air used in Germany for assessing exposures is the "Allgemeiner Staubgrenzwert" ("general threshold limit value for dust"), which is 3 mg/m^3 and 10 mg/m^3 [7] for the respirable and inhalable dust fractions, respectively, and applies to poorly soluble and insoluble dusts[1]. For the respirable dust fraction, there is a limit value in air of 6 mg/m^3 for a number of separately specified working areas [8].

Frequently there are also substance-specific limit values in air for certain components in the dust fractions [7]. These must be complied with independently of the German "general threshold limit value for dust".

The "general threshold limit value for dust" and substance-specific limit values are defined as shift average values. For short-term higher exposures, short-term exposure limit values were introduced which are to be monitored as 15-minute mean values. In the majority of cases, excursions of up to 4 or 8 times the limit values in air are currently permitted [4, 6, 7]. The permitted total duration of increased exposure is 60 minutes per shift. Therefore, depending on the occupational activity to be monitored, it is always necessary to check whether the shift average value and/or the short-term exposure limit needs to be measured.

The following describes the monitoring of limit values in air for the above-mentioned dust fractions and their chemical components. The main focus will be on the discontinuous measuring procedures. Samples are collected on site and subsequently processed in the laboratory. This is usually achieved by gravimetrical determination of dust mass. Determination of individual chemical components of dust may require sample preparation.

Direct-reading methods for dust will not be discussed here. However, a number of devices already exist which allow on-site determination of dust exposure at workplaces [9]. Similarly, certain substances (e. g. diesel particulate matter [10], welding fume, quartz [11]) will not be discussed because special monitoring requirements generally need to be met in these cases.

2 Performing dust measurements

A dust measurement essentially comprises the following working steps:

- selection and preparation of sampling devices,
- strategic sampling,
- sample collection,
- sample preparation and analysis.

---

1 This definition for the assessment of workplaces is based on the MAK Commission's threshold limits for "A" and "E" (respirable and inhalable) dusts, which are 1.5 mg/m^3 and 4 mg/m^3, respectively.

Sampling for the purpose of determining dust exposure in the workplace is always conducted actively by means of sampling pumps. Passive sampling procedures which collect dusts by sedimentation, as in the environmental area, are not employed in workplace measurements.

Sampling is carried out using sampling heads which contain the medium for the collection of dust. These sampling heads are connected to a sampling pump, which then draws the workplace air through the system.

When carrying out workplace measurements it needs to be ensured that measurement uncertainty meets the definitions in EN 482 [12]. This means that the contributions to inaccuracy from each of the above-mentioned working steps must be kept as small as possible. This is achieved by using sampling devices (sampling pump, sampling head and collection medium) and analytical procedures that meet the given requirements. Before selecting the sampling time it should be checked, by comparison with the limit of quantification, whether monitoring of the limit value in air is feasible.

2.1 Requirements for sampling devices

2.1.1 Sampling pumps

Sample collection can be performed as personal or stationary sampling. This constitutes the first major reason for differences in pump performance. In personal sampling, a compromise has to be achieved between displaced flow rate and use by the wearer. The wearer must not be hindered in his or her activities. Sampling pumps in stationary sampling systems are not subject to the latter limitation, so that, as a rule, pumps are used whose displacement performance is 1–2 orders of decimal magnitude greater. The limitations of personal sampling are mainly due to battery capacity, which is clearly limited by the size of such pumps.

The performance of personal sampling systems is tested in accordance with EN 13205 [13].

Personal sampling

Sampling pumps with volumetric flow rates in the range of 5 mL/min to 5 L/min which are used for personal sampling must meet the requirements of European Standard EN 1232 [14]. Furthermore, only type P sampling pumps, which are intended for personal sampling of dusts, will be discussed here. Personal sampling pumps with flow rates greater than 5 L/min must fulfil the requirements of European Standard EN 12919 [15].

The pumps used must be equipped with either automatic flow control, which maintains a constant volumetric flow rate in the presence of varying flow resistance, or with a device for determining the sampling volume.

For use in areas where there is a danger of explosion, the requirements laid down in EN 50014 [16] must be satisfied.

Pumps meeting the requirements listed in Table 2 can be used for personal sampling of the respirable and inhalable dust fractions. It is important to ensure that the pump's flow rate is always adjusted to the sampling pump/sampling head/collection medium system. The set flow rate should be rechecked after sampling.

Table 2. Overview of the Standards requirements for sampling pumps for personal sampling [14, 15].

Parameter	Requirement	Note
Mass	< 1.2 kg	Not applicable to pumps with flow rates > 5 L/min
Mechanical strength	≤ 5% deviation in flow rate from the initial value	After a shock treatment (general functioning of the pump must be unaffected; there must be no mechanical or electrical defects)
Pulsation of flow rate	≤ 10%	
Flow rate stability under increasing back pressure	≤ 5% of the initial value	Applies to any measured back pressure-flow plot (typical of unloaded and heavily loaded filters)
Duration of operation	Preferably 8 hours, but at least 2 hours	In battery operation; applies to the entire nominal flow rate range
Start and long term behaviour	Volumetric flow rate must not deviate from the initial value by more than 5%	Operation at room temperature and at 5 °C
Short term reduction of flow rate	Shutdown or activation of malfunction indicator	Complete interruption of volumetric flow rate for 2 min
Temperature dependence	≤ 5% deviation in flow rate from the value at 20 °C	At ambient temperatures from 5 °C to 40 °C
Orientation	≤ 5% deviation in flow rate from the value for the upright position	When the pump is tilted by 90° in any direction; not applicable to stationary pumps
Timer accuracy	≤ 5 min deviation from the reference timer after 8 hours	Applies only when equipped with an internal timer

Appendix 1 contains a selection of currently available pumps for personal sampling of the respirable and inhalable dust fractions.

Stationary sampling

Pumps used in stationary sampling operate at flow rates which exceed by about 1–2 orders of decimal magnitude the flow rates used in personal sampling. As a result, markedly lower concentrations can be measured, or sampling time can be reduced to such an extent that peak exposures can be monitored.

Pumps for stationary sampling must also meet the Standards requirements listed in Table 2. Exceptions exist only in respect of mass, on which there are no restrictions because use is stationary, and orientation.

The sampling systems currently on the market frequently form a unit consisting of a sampling pump and a sampling head (see Section 2.1.2). A list of these systems is provided in Appendix 2.

2.1.2 Sampling heads

There are various commercially available sampling heads for stationary sampling and personal sampling of the "A" and "E" (respirable and inhalable) dust fractions. Meanwhile devices are also available for simultaneous collection of the "A" and "E" dust fractions on the wearer. Appendix 2 provides an overview of currently available sampling heads.

A pump is used to draw the air through the sampling head. The suction flow rate at the inlet of the inhalable dust sampling head shall be 1.25 m/s (± 10 %). By utilising various physical principles of separation, the desired dust fraction is then collected on the filter in the sampling head. The stationary sampling heads are equipped with a matching pump which is part of the sampling system and designed for a specific sampling head. Personal sampling heads should be used with the pumps listed in Appendix 1.

Some of the sampling heads listed in Appendix 2, when combined with a suitable collection medium, can be used for simultaneous sampling of particles and solvent vapours. These systems are used e. g. in the sampling of lacquer aerosols [17].

2.1.3 Collection media

Workplace dust sampling is predominantly performed with filters – and more recently also polyurethane foams – of different pore sizes as collection media. The (filter) materials employed should be matched to the conditions at the sampling site but also to the analytical requirements. Appendix 3 provides an overview of frequently used filter materials and collection media.

Thermal stability, dust adhesion and high loading capacity in combination with low differential pressure of the filter element (filter resistance) play an important role at the sampling site. High mechanical strength and a minimum of electrostatic charging at weighing are essential factors for the suitability of a filter material for dust sampling. If, in addition to dust weight, chemical components of dust (e. g. PAHs, metals) are to be determined, further parameters need to be taken into account. For instance, filter blank values, matrix effects, resistance of the filter material to acids and solvents or wetting properties may affect the subsequent analytical process (see also Section 7).

Generally, the filters used should have a separation efficiency for paraffin oil mist [18] of at least 99.5 % [19].

Frequently, the various requirements that optimally suitable filter material must meet for sampling are not compatible with those for the subsequent analytical process and, hence, a compromise needs to be found which does justice to both metrology and analysis.

The characteristic properties of frequently used filter materials are described in the following.

2.1.3.1 Glass fibre filters

Glass fibre filters are depth filters. They consist of compressed glass fibres forming an ir-regular three-dimensional network with many interspaces of various sizes. Hence, parti-cles are not only retained on the surface but also trapped deep within the filter structure.
Glass fibre filters feature high collection capacity for dusts in combination with good adhesion properties. They tolerate high flow rates and offer good wet strength.
Due to the large surface area of filters, compounds of higher volatility (e.g. some PAHs or organometallic compounds) contained in the dust evaporate from the filter during prolonged storage.
Another disadvantage is that the high and often variable blank values of this filter mate-rial can affect the analysis of metals in dust. These filters are unsuitable, or only poorly suitable, for instance for the determination of zinc and iron (unfavourable limits of quantification). Similarly, the determination of aluminium, calcium and barium also creates problems. Even the now available binder-free, low-blank glass fibre filters made of borosilicate glass are poorly suited for use with sensitive analytical methods (ET-AAS, ICP-MS).
As a rule, only binder-free glass fibre filters that are low in trace elements should be used, because high levels of binders could lead to false-positive results, e.g. in the ana-lysis of diesel particulate matter [10].
Their excellent mechanical properties and thermal stability make these filters ideal for gravimetric analysis of airborne dust.

2.1.3.2 Quartz fibre filters

Quartz fibre filters (correct term: quartz glass fibre filters) are also depth filters and their sampling properties are similarly as good as those of glass fibre filters.
They are considerably better suited for the analysis of metal components in dusts due to their extremely low and relatively constant blank values. However, the presence of soluble silicates may, again, persistently interfere with analysis when the AAS graphite furnace technique is used.
Some brands of quartz fibre filters are mechanically sensitive so that problems can arise during gravimetric analysis.

2.1.3.3 Membrane filters

Membrane filters are surface filters because the dust is collected exclusively on the top of the membrane. Their extremely porous surface makes the filters very light. They usually consist of cellulose nitrate or acetate or are manufactured from mixed cellulose esters. Their loading capacity is markedly lower than that of depth filters. Hence, they are less suitable for high air throughputs at high dust levels. Problems may also arise when sampling for very fine particles (e.g. welding fumes). Furthermore, it is important to remember that the filters are combustible and must be protected against flying sparks.
Filters with pore sizes of 0.8–8 µm are frequently used. It should be borne in mind that particles smaller than the absolute pore size are also retained due to certain mechan-isms in the membrane.

The usually hydrophobic properties of the membranes may also have disadvantageous effects when the filters are used in high-humidity areas (e.g. plating shops). However, specifically hydrophilic membrane filters are also commercially available.

Membrane filters are very well suited for the subsequent analyses for metals contained in dust because their blank values are near zero and they are readily digestible owing to their low resistance to acids.

Nevertheless, blank values also cannot be neglected in the case of membrane filters. In very sensitive analytical methods (e.g. ET-AAS), residual nickel, for instance, may interfere with the detection of low concentrations. These blank values are attributable to the use of natural, already contaminated cellulose in the manufacture of the filter materials. Regular checks of blank values are, therefore, also indispensable for membrane filters.

If membrane filters are used in gravimetric analysis, various interfering factors can impede analysis (see Section 6).

2.1.3.4 Fluoroplastic (Teflon) filters

These filters are manufactured from polytetrafluoroethylene (and also polypropylene for backing layers). The membrane is extremely resistant to acids, alkalis and solvents. The filters are of very high thermal stability and exhibit hydrophobic properties.

Ideally, these filters are used in working areas in which aggressive, acidic aerosols must be reckoned to occur. Their very low blank values enable high-sensitivity determinations. The filters are very readily extractable by sonication and are used e.g. in the analysis of semi-volatile PAHs.

The filters are basically suitable for gravimetric analyses but expensive and therefore only used in exceptional cases requiring high precision.

2.1.3.5 Polyurethane foams

In addition to filters, polyurethane foams (PU foams) are also employed as collection media [20]. These specially purified PU foams permit the sampling of both gases (e.g. volatile dioxins/furans) and dusts. Most of them are open-cell TDI polyether-based flexible foams of different, controlled porosities. This is where the great advantage lies over the filters described above. The choice of polyurethane foams of various densities/pore sizes allows the collection and separate analysis of different dust fractions in one sampling. No special sampling heads (e.g. cyclones) are required.

The sampling behaviour of polyurethane foams is similar to that of depth filters. The dust is deposited on the surface of, and in the spaces within, the PU foam. By virtue of its large, polar surface, the material exhibits a high collection capacity for dusts.

Gravimetric analyses are possible but dust adhesion leaves much to be desired if the foam pore size is large.

It remains to be investigated more thoroughly whether the material is also suitable for the analysis of metals. PU foams can contain many production-related impurities from auxiliary components (e.g. organotin compounds), and therefore prepurification of the material is crucial in order to reduce blank values. In addition, the material's moderate solubility can also interfere with its digestion.

Furthermore, when using multi-stage sampling systems, the separation of dust into different fractions may give arise to problems regarding the limit of quantification and the interpretation of the measurement results [21, 22].

3 Sampling strategy

The basic requirements for the determination and assessment of workplace concentrations of hazardous substances in Germany are laid down in "Technische Regeln für Gefahrstoffe (TRGS; Technical Rules for Hazardous Substances) 402" [6] and DIN EN 689 [23].

The aim of measurements in workplace air is to obtain as comprehensive and representative an assessment as possible of an exposure situation in a defined working area. The working area may be defined in terms of a spatial area or employees' working activities during a shift. The working area under assessment has to be precisely defined and described as part of a purpose-specific analysis.

Implementation of the procedure as recommended by TRGS 402 is to ensure an objective assessment of the particular working area by comparing the measured concentrations of hazardous substances in the workplace air with the relevant occupational exposure limit in air. Before sampling can be started, comprehensive investigations taking into account the chosen measuring strategy are necessary to enable selection of a suitable measuring procedure (sampling and a corresponding analytical method). Sampling conditions need to be defined and customised for the measuring task according to the occurring substances and the expected concentration ranges.

In order to obtain a result for the occupational exposure assessment, the concentration of a hazardous substance needs to be compared, in air in the form of a shift average value, with the relevant occupational exposure limit in air. For the determination of a shift mean value, there are requirements on sampling conditions (site and duration of measurement).

3.1 Type of measurement

A distinction is made between several types of measurements which are associated with a specific sampling strategy. These are:

– exposure measurements (shift average, short-term and monitoring measurements),
– worst-case measurements,
– measurements at the emission source,
– other measurements.

When reporting, it is necessary in each case to indicate the type of measurement.

Exposure measurements

Exposure measurements are measurements of hazardous substance concentrations in workplace air that detect one or several components. They do not allow stringent conclusions about shift average values. They are also used in the process of occupational exposure assessment to obtain preliminary information.

Shift average values

To enable the shift average value to be obtained from exposure measurements, statistical methods were used to create Table 3, which specifies the necessary duration of sampling and the corresponding minimum number of samples that need to be taken during a shift. Only such sampling durations for dust sampling were included as actually enable representative assessment of exposures in practice.

Table 3. Duration of sampling and minimum number of samples [6].

Duration of sampling (averaging period)	Minimum number of samples
15 min	≥ 4
30 min	≥ 3
1 hour	≥ 2
≥ 2 hours	≥ 1

Before selecting the sampling time it should be checked, by comparison with the limit of quantification, whether monitoring of the limit value in air is feasible.
In the event that several measured values were obtained in the course of a shift, a time-weighted shift average value (C_s) is calculated according to Equation (1):

$$c_s = \frac{\sum c_i \cdot t_i}{\sum t_i} \tag{1}$$

where:
c_i is the concentration of substance i in the workplace air and
t_i is the sampling duration for measured value c_i

Exposure-free periods during a shift, if they occur, must also be included in the calculation of the shift average value according to Equation (1) ($c_i = 0$). In addition, the conditions for short-term exposure limit values should be observed, which in the worst case result in excursions above the limit value even though the calculated shift average value is below the limit value. During sampling, breaks from work and other times at which the working area is left need to be accounted for by switching off the sampling pump.

Short-term exposure measurements

The aim of these measurements is to detect short-term increases in exposure. The excursion factor has been set at 1 for a number of chemical components of dust, i.e. the limit value must not be exceeded at any time. Numerous insoluble, non-toxic dusts and most toxic metal dusts have been assigned an excursion factor of 4. Periods of short-term increases in concentration are usually averaged over 15 minutes. This requirement places special demands on sampling technology, the selection of sampling periods during the shift and, not least, analytical methodology. The planning of measurements requires detailed knowledge of the flow of work and the technology used. It must be ascertained when, where, how often and for how long peak exposures occur. Measurement planning has to take this into account and, whenever possible, consider continuous measuring procedures in order to obtain the required data covering appropriate measurement periods. However, continuous measurements are possible only in a small number of cases, and therefore, as a rule, cumulative 15-minute sampling has to be performed. Should the analytical method not be sufficiently powerful, then several short-term collection periods should to be accumulated on one collection medium.

Sampling requires a sufficiently high volumetric flow rate in order to enable the assessment of the criteria for short-term exposure limits for respirable or inhalable dust because the limit of quantification in gravimetric filter analysis constitutes the limiting factor. New personal sampling systems with flow rates of up to 10 l/min, the use of stationary sampling systems with high volumetric flow rates and the combination of gravimetric analysis with a direct-reading dust monitor offer metrological possibilities to detect and quantify short-term exposures.

Monitoring measurements

Monitoring measurements serve the purpose of checking the observance of limit values after occupational exposure assessment has been completed. The time interval between the monitoring measurements depends on the outcome of the occupational exposure assessment. The target analytes (e.g. an indicator component), measuring procedure and sampling conditions will have been defined in the context of the occupational exposure assessment.

Intervals for monitoring measurements, as shown in Table 4, depend on the ratio between the measured shift average value and the limit value (substance index). For mixtures of substances, the shift average values for the individual substances are used to calculate a substance index, and the sum of the substance indices (exposure index) is then compared with 1 [24].

Table 4. Schedule for monitoring measurements.

Exposure index	Interval between monitoring measurements
$I \leq 0.25$	≤ 64 weeks
$0.25 < I \leq 0.5$	≤ 32 weeks
$0.5 < I \leq 1$	≤ 16 weeks

Worst-case measurements

Worst-case measurements are performed in situations where markedly increased exposures are expected due to special exposure conditions. This may be the case e.g. when highly dust-producing substances are used or when incoming orders make it necessary to increase throughput to the limit of capacity. If observance of the limit value is demonstrated under such conditions it can be assumed that the limit value in air is also observed under the usual conditions.

Measurements at the emission source

Measurements at the emission source are based on the same considerations as worst-case measurements. Similarly, observance of the limit value in the immediate vicinity of an emission source can be extrapolated to the entire working area, allowing a reduction in measurement effort.

Other measurements

Data on specific individual activities, work flow, working techniques as well as individual substances can be obtained by performing specific measurements which are not intended for the assessment of an entire working area or the calculation of a shift average value. Such limited findings can, however, be quite necessary and valuable in determining the state of the art or in the context of preliminary studies.

3.2 Type of sampling

A distinction is made between personal sampling and stationary sampling. Since the objective is to assess the exposure of employees, personal sampling should be given preference. Depending on the type of measurement, however, stationary measurement may also be indicated, as it does not, as a rule, adversely affect or interfere with the normal work flow of the monitored employees in a working area. Stationary sampling can, for instance, readily be used for measurements at an emission source. The disadvantages of stationary sampling lie mainly in the fact that the results deviate from personal measurements carried out at the same time, because

- employees do not remain in the same place for the entire duration of sampling,
- the stationary sampling device can be obstructed by the employee,
- the concentration depends on the distance from the emission source, and
- the local concentration is influenced by air currents.

Stationary sampling should be conducted at the employee's breathing level; this is usually at a height of approx. 1.5 m.
Depending on the positioning of the sampling device, findings can be higher or lower compared with personal sampling.
Measured values obtained by personal sampling may be too low if the working activities allow the sampling head to become covered. Therefore, right-handers should have the sampling head fixed on the left, and left-handers on the right-hand side of the body.

4 Sampling

4.1 Recording the sampling conditions

Measurements of hazardous substance concentrations in workplaces are usually considered only under the aspect of assessing current exposure conditions. However, they also serve as long-term documentation of the exposure situation in a workplace, so as to provide a reliable database in the event of any occupational diseases which may occur at a later time.

For these reasons it is important that the sampling conditions be documented in a comprehensive and comprehensible manner [6, 25].

The following types of information related to sampling conditions are indispensable.

- List of working substances
  All substances used in the workplace and all substances (or groups of substances) occurring in (released into) the workplace air should be listed. In this context, dermal absorption should also be taken into account. If measurement is limited to specific hazardous substances, reasons must be given to explain why other hazardous substances are not relevant or to what extent they can be included in the assessment.

- Description of the working area
  - definition of the working area:
    in terms of space, organisation or activities
  - location of workplaces:
    e.g. outdoors, in closed areas, partly open
  - description, type and/or purpose of the working procedure
  - connections with further, in particular adjacent, working areas as far as they affect exposure
  - spatial conditions (ground plan), drawings, images
  - area and height of the premises
  - details of windows, doors or other areas via which air can be exchanged

- Type of machinery and working materials
  - type, year of construction, number, production conditions, working method
  - power rating, production parameters, operating times, emission sources

- Description of the working procedure
  - working process, times of exposure, working hours and peak exposures
  - number of employees/number of workplaces in the working area
  - personal protective measures/equipment
  - organisational preventive measures

- Technical protective measures
  - measures against emissions:
    e.g. process engineering measures, hazard detection technology, encapsulation
  - ventilation
  - air flow control

- Sampling procedures
  - stationary or personal
  - dust fraction to be analysed (inhalable or respirable)
  - sampling equipment used (sampling head, filter material and sampling pump)
- Sampling
  - assignment of samples to a place/employee
  - duration of sampling and volumetric air flow
  - Ambient data\environmental data (indoors and outdoors)
  - description of working and operating conditions at the time of measurement

4.2 Possible sampling errors

Experience shows that in the measurement of hazardous substance concentrations in workplace air, sampling makes the main contribution to measurement error. TRGS 402 requires that measurement uncertainty, as the total error encompassing all systematic and random errors that occur during measurement, not exceed 30%.
Apart from methodical errors (sampling site/time, failure to measure all relevant substances, wrong strategy), there are a number of other possible sampling-related errors. When sampling, the following points should therefore be checked:

- Leaks in the system during sampling (pumps, tubing, filter holder, sampling head).
- Contamination with residues in the sampling heads (maintenance and cleaning of the sampling devices).
- Maintenance and calibration of the pumps (stability of the set flow rate, capacity of the battery, accuracy of time measurement).
- Electrostatic charging (e.g. of the cellulose filters or the sampling head).
- Consideration of extraneous deposits on the filter or rupture of the filter material due to the entry of coarse-disperse particles resulting from directed movement (e.g. when using a cutting disk, sawing or welding).
- Contamination during transport.

The errors listed above can, as a rule, be avoided by technically appropriate and correct selection, handling, cleaning and maintenance of the sampling device.
Even if all known possibilities of error are taken into consideration, every measured value will, of necessity, include a measurement error. The EN ISO/IEC 17025 standard [26] therefore obliges the user to determine measurement uncertainty.
Exact determination of the sampling volume is of particular importance. This is due to the characteristics of the sampling pumps (see Section 2.1) on the one hand and the accuracy of the measurement of flow rate on the other. Errors well below 2% can be achieved with drum gas meters. Mass flow meters and soap-bubble flow meter are also well suited; rotating ball flow meters can exhibit errors of more than 5%.
Limit values in air refer to a temperature of 20 °C and an atmospheric pressure of 1013 hPa. The German "Arbeitsstättenverordnung" (Workplaces Ordinance) [27] makes provisions for ambient temperatures between 12 °C and 26 °C, depending on the heaviness of work. In practice, however, temperatures occur that are higher or lower than

these values. The atmospheric pressure extremes in Germany are 955 and 1058 hPa. Deviations from standard temperature and pressure conditions are small as a rule. In the worst, deviations in sampling volume can be approx. 10%, e.g. in underground mining. Therefore the sampling volume should be corrected for pressure and temperature, as appropriate, in order to limit total measurement error. It should also be taken into account that the set flow rate of the sampling pumps can also be affected by changes in environmental conditions during sampling (e.g. by an increase in temperature during the shift).

5 Treatment of samples

5.1 Transport

The filters are transported in filter holders sealed with dust-proof caps or stoppers together with, if required, the blank filter or the sets of blank filters. Transport should take place as shock-free as possible. Filter cassettes should be sealed such that they cannot open by themselves (e.g. by shocks or vibrations). Special cooling is required only for a small number of dusts which contain volatile components, e.g. polycyclic aromatic hydrocarbons (PAHs).
The filter holder/cap combination should be such that, in the event of the dust spreading within it, recovery to the filter should be possible without significant loss, such recovery being necessary if gravimetrical determination of dust mass is to be performed (exception: the filter, filter holder and cap are weighed together, which is a possibility according to ISO 15767 [28]). The spreading of dust in the holder is often observed with inhalable dusts in particular as they adhere less well to the filter than do the finer respirable dusts.
The material for the filter capsules and cassettes should be selected such that static charges, which can result in the adherence of dust within the sealed container, are avoided as best possible.

5.2 Storage of loaded filters

The storage of loaded filters depends on the type of analytes to be determined. Generally, filters for gravimetrical determination of dust mass according to Section 6 are treated using one of the following methods, depending on the organisation of work flows in the laboratory and the nature of the samples.

a) *Individual labelling of the sample on the filter*: The filters are removed from the holder with a pair of tweezers (if necessary after transferring onto the filter the dust distributed in the holder, e.g. by means of a fine brush) and placed in clean Petri dishes in such a manner that mutual contamination is not possible. Sample identification can be recorded on the Petri dish using a permanent marker. When using cel-

lulose nitrate or cellulose acetate filters or mixed ester filters, an identification number is noted on the rim of the unloaded filter with a ballpoint pen before weighing,
the rim remaining free of dust when the filter is mounted in the holder. The Petri
dishes are covered loosely with a glass top for dust protection.

Alternatively, the filters can be stored on a rack for thin-layer chromatography
plates, which can be placed in a desiccator [10]. This offers the advantage that a larger number of filters can be stored in a space-saving manner and, if necessary, with
a desiccant.

b) *Individual labelling of the sample on the filter holder/cassette (e.g. bar code):* The
filters are kept vibration-free in the cassettes or capsules until conditioning for re-
weighing.

The storage time before weighing should not exceed 4 weeks.

6 Determination of the dust concentration

As a rule, the dust mass on the filters is determined by weighing them. There are various
ways in which interferences (relative humidity, transport and sample handling) can be
taken into account when determining the method performance parameters. The following
describes three methods elaborated on the basis of the requirements of ISO 15767 [28].

Method 1:   Use of laboratory blank filters (see Section 6.4.1)
Method 2:   Calibration using environmental data (see Section 6.4.2)
Method 3:   Inclusion of blank filters in the sampling procedure (see Section 6.4.3)

It is important to ensure that a series of measurements is performed only with filters
from the same production batch.

In addition to weight determination, sample carriers for the VC 25 sampling system
also allow the concentration of the respirable dust fraction to be determined by means
of β-absorption (attenuation of β-radiation by the dust loadings on the filters) [29].

6.1 Equipment and materials

Weighing requires the following items of equipment and materials:

Analytical balance: Readability 0.01 mg; reproducibility 0.02 mg; equipped to re-
                         duce electrostatic charging; weighing pan adapted to the
                         weighing task (see Section 6.3)

Petri dishes: Glass, 150 mm diameter (sample storage according to Section
                         5.2 a)

Storage boxes: Dust-free, covered boxes for storing filters in filter holders/
                         cassettes (storage of samples according to Section 5.2 b)

Tweezers:	18/8 stainless steel
Brush or rubber scraper:	For transferring dust from the filter holder/cassette to the filter
Air ioniser:	e.g. Eltex 108EK360 (ELTEX-Elektrostatik GmbH, D-79576 Weil am Rhein) Ionising bar (Haug GmbH & Co. KG Ionisationstechnik, Leinfelden-Echterdingen)
Ambient air monitoring:	Devices for recording relative humidity and temperature, e.g. the Almemo 2290–3 environmental data logger, Ahlborn Mess- und Regeltechnik, Holzkirchen (recommended; required for Method 2, see Section 6.4.2)
Bar code reading:	Scan code and bar code reader and decoder (e.g. from CS Computer Systeme GmbH, Baierbrunn) (optional)
Data acquisition:	For example: personal computer for the capture of the weighing data and, where applicable, the environmental data

6.2 Conditioning the filters

The filters – unloaded for weighing and dust-loaded for re-weighing – are conditioned (e.g. in Petri dishes) in the weighing room for at least 6 hours prior to weighing. To this end, the filters are stored in a manner allowing them to equilibrate in the laboratory atmosphere, i.e. access of air must be ensured. When storing filters in filter cassettes after loading, this is achieved by removing the lower part of the cassette and placing the cap on the holder in a slightly tilted position. The opened cassettes must, in this case, be placed in a dust-free holder that is closed at the top.

6.3 Weighing the filters

Depending on the purpose of use, filters composed of e.g. cellulose nitrate, glass fibres, quartz glass fibres, PVC, mixed esters, Teflon or PU foam are weighed.
It is advantageous to use a balance with anti-static equipment (grounded perforated-metal weighing pan with a diameter >150 mm; draft shield with grounded metallised glass surfaces). The filter is placed on the weighing pan of the analytical balance using a pair of tweezers. When using filters with a tendency to build up static charge (e.g. cellulose nitrate membrane filters), the reverse side of the filter should, if necessary, be treated briefly with an air ioniser (fan switched off!) from a distance of approx. 10 cm, or the filter should be passed over an ionising bar while being introduced into the balance weighing chamber in order to neutralise electrostatic charges. As soon as the weight has reached a constant value, the reading is recorded or captured electronically.
It should be checked before re-weighing loaded filters, whether any dust has become dislodged from the filter or the filter holder/cassette contains loose dust as a result of high loading. This is found more often in determinations of "E" (inhalable) dust than

in determinations of "A" (respirable) dust. The loose dust must be transferred to the filter (e.g. with a brush) after its removal from the holder. If it is found that a significant amount of the dust cannot be transferred to the filter, or if there are indications that part of the dust loading was lost during transport after sampling, a note of this should be made as an amendment to the weighing result (e.g. "Sample loss may have occurred during sample treatment or transport; measured value = minimum value"). The same procedure should be used if the filter has been damaged to the extent that parts of it are missing (e.g. if the rim of the filter has been torn off because it was mounted in the sampling head too firmly). In this case, it is not meaningful to determine dust weight by weighing. Any manipulations of the filter (e.g. wiping traces) should also be indicated.

Depending on filter size, it is recommendable to have several different weighing pans or, in the case of small filters, to use a fixture for the weighing pan that will facilitate the handling of small filters.

The balance should be tared at regular intervals or after a predefined number of weighing operations. If the balance takes unusually long to reach a constant weight reading, or even exhibits a drift, any static charge that may be present on the balance should be neutralised with an air ioniser/ionising bar prior to further use.[2]

6.4 Determining the characteristics of the procedure and the dust concentration

The calculation of the dust concentration and the determination of the limit of detection depend on the method of weighing. The relevant procedural steps for the three methods are described in the following.

6.4.1 Determination of the dust concentration, taking into account the effects of environmental conditions by using laboratory blank filters (Method 1)

Brief description of the method

The method is used for filters made of e.g. cellulose nitrate, glass fibres, quartz glass fibres, PVC, mixed esters, Teflon or PU foam. In addition to the measurement filter, three reference filters of the same type and size are both weighed and re-weighed. The effect of the environmental conditions on the weight of the measurement filter is taken into account by introducing into the calculation the relative weight change of the reference filters at the time of re-weighing.

---

2 It is recommended, regardless of which of the methods is used as the weighing procedure, to check the environmental data (relative humidity and temperature). Even air-conditioned rooms – depending on the type and capacity of the air conditioning system – are subject to inevitable variations in temperature and relative humidity (usually about ± 2 °C and ± 5 % relative humidity). This must be taken into account especially after short-term changes in weather (e.g. thunderstorms). Weighing operations should not be performed if the environmental data exhibit a marked drift.

Weighing procedure

Three reference filters of the same type as the measurement filters must be weighed immediately before or after those filters themselves are weighed. There is a set of such reference filters for every type and size of filter. Once a year, a new set of reference filters is established for each type of sample carrier. After sampling, the loaded filters are also re-weighed versus the set of reference filters. In this way, the relative humidity of the blank value for the measurement filter is corrected via the weight change of the reference filters at the time of re-weighing (for calculation, see Equation (2)).

Temperature, humidity and atmospheric pressure are recorded for reference filter weighing.

Calculation of the dust concentration

The calculation of dust weight (net weight) on the filter, corrected for humidity, is performed according to Equation (2).

$$X = m_r - (m_p \times m_{rr}/m_{rp}) \tag{2}$$

where:

m_{rp} is the mean weight, in mg, of the three reference filters at the time the measurement filters are pre-weighed

m_{rr} is the mean weight, in mg, of the three reference filters at the time the measurement filters are re-weighed

m_p is the weight, in mg, of the measurement filter prior to sampling (measurement filter weight at pre-weighing)

m_r is the weight, in mg, of the measurement filter after sampling (measurement filter weight at re-weighing)

X is the dust weight on the filter, in mg.

Limit of detection and limit of quantification

The limit of detection (LOD) is three times the standard deviation of the weight difference (weights determined before and after shipment) for a minimum of ten unloaded filters having undergone the complete procedure, including transport to the measurement site and back within a period of time that is usual for such procedures. The limits of quantification (LOQ) are estimated as ten times the standard deviation. Table 5 gives typical limits of quantification for frequently used types of filters as obtained by the laboratory blank filter method in a weighing room without air conditioning.

6.4.2 Determination of the dust concentration, taking into account the effects of environmental conditions by calibration with environmental data (Method 2)

Brief description of the method

The method has so far been used for cellulose nitrate and glass fibre filters as well as PU foams. Test series are run for each type and size of filter before measurement filters are weighed to determine the dependence of the empty filter weights on environmental conditions (significant only for relative humidity). The weighing results for the measurement filters are corrected to the nominal value for relative humidity by using a formula which incorporates relative humidity at the time of weighing. Glass fibre filters exhibit no significant weight differences at relative humidities between 30 % and 70 %.

Determination of parameters for calibrating the weight of measurement filters

Per type and size of filter, a minimum of three filters from one batch are used to determine the calibration data. The filters are stored separately in Petri dishes (conditioning according to Section 6.2). A series of weighings is performed at different time intervals over the course of several weeks. In this process, weighing results are obtained under different environmental conditions. The range of environmental data should cover at least the variations usually observed. However, it is recommendable additionally to set the air conditioning settings to higher and lower relative humidity values (e.g. 30 % and 65 %) for a limited period of time, so as also to carry out weighings under those conditions. A minimum of 20 weighing results should be obtained which permit a linear regression over the relevant environmental data range. This calculation yields the change in filter weight per percent of change in relative humidity. The result is compared for the differently sized filters made of the same material. In the absence of outliers, the mean change in filter weight per percent change in relative humidity is adopted as the correction factor. In small filters (diameter ≤ 37 mm), however, the effect of relative humidity on filter weight is within the range of filter weight variability. For practical application, a correction factor should therefore be chosen that was determined using larger filters of the same material.

It is not necessary, at least for air-conditioned laboratories, to convert relative to absolute humidity. The effect of temperature within the typical range of temperature variability ($\pm 2\,°C$) is negligible for the correction of the weighing results.

Calculation of the dust concentration

Equation (3) is used for approximate conversion of the determined filter weights (for empty and loaded filters) to values at 50 % relative humidity.

$$X = m_w + m_w \times (50 - F) \times f_c \qquad (3)$$

where:
X is the weighing result, in mg, corrected to 50 % relative humidity
m_w is the weighing result, in mg
F is the measured value of relative humidity, in %
f_c is the correction factor (relative change in filter weight at a difference
    in relative humidity of 1 %, as determined by linear regression), in $\%^{-1}$

The dust concentration ρ_d is calculated according to Equation (4):

$$\rho_d = (m_{cl} - m_{ce})/V \qquad (4)$$

where:

m_{cl} is the corrected weighing result, in mg, for the loaded filter
m_{ce} is the corrected weighing result, in mg, for the empty filter
V is the sample volume, in m^3.

Limit of detection and limit of quantification

The limit of detection (LOD) is three times the standard deviation of the weight difference (weights determined before and after shipment) for a minimum of ten unloaded filters having undergone the complete procedure, including transport to the measurement site and back within a period of time that is usual for such procedures. The limits of quantification (LOQ) are estimated as ten times the standard deviation. Table 5 in Section 6.5 provides, for several frequently used dust sampling filters, an overview of typical limits of detection as determined by the present method.

6.4.3 Determination of the dust concentration, taking into account the effects of environmental conditions by including blank filters in the sampling procedure (Method 3)

Brief description of the method

Correction for possible effects on sampling, sample transport and storage is achieved using the mean weight difference of blank filters included in the sampling process. One blank filter per ten measurement filters should be included for this purpose. There should be a minimum of five blank filters per measurement series. If the number of filters that was prepared but not used for sampling is greater, all unloaded filters are used for the determination of the blank value. This procedure can be used for glass or quartz fibre filters and for cellulose nitrate or acetate filters.

Weighing procedure

Before sampling, the unloaded filters and, after sampling, the loaded and unloaded filters are separately conditioned over silica gel (containing a humidity indicator) in a desiccator for at least twelve hours. The filters are removed from the desiccator half an hour before weighing, and stored next to the balance. The filters are then weighed. Here it is important to check for weight constancy.

Limit of detection and limit of quantification

The filters not loaded during sampling serve as blank filters for the determination of the parameters.
The unloaded filters are weighed before and after sampling. The weight difference of all unloaded filters as calculated according to Equation (5) is used to determine the limit of detection.

$$m_{bi} = m_{bai} - m_{bbi} \tag{5}$$

where:
m_{bi} is the weight difference, in mg, of the unloaded filters before and after sampling
m_{bbi} is the weight, in mg, of the unloaded filter before sampling
m_{bai} is the weight, in mg, of the unloaded filter after sampling

The weight differences determined for the unloaded filters are tested for outliers and any identified outliers removed from the data set.
Subsequently, the mean value and standard deviation of the weight differences are determined for the unloaded filters.

$$m_b = \sum m_{bi}/n \quad (i = 1, \ldots n) \tag{6}$$

and

$$s_b = \sqrt{\sum(m_b - m_{bi})^2/(n-1)} \tag{7}$$

where:
m_b is the mean weight difference, in mg, for the unloaded filters (blank value)
s_b is the standard deviation from the mean weight difference, in mg, of the unloaded filters

The limit of detection for loaded filters is three times the standard deviation of the unloaded filters $(3 \times s_b)$.

Calculation of the dust concentration

For a loaded filter, the loading is determined by a procedure analogous to that for the blank filters. The blank value (m_b) is subtracted from the determined weight.

$$X = (m_r - m_p) - m_b \tag{8}$$

where:
m_p is the weight, in mg, of the loaded filter prior to sampling
m_r is the weight, in mg, of the loaded filter after sampling
X is the dust weight on the filter, in mg.

The dust concentration ρ_d is then calculated according to Equation (9).

$$\rho_d = X/V \tag{9}$$

where:
V is the sample volume, in m^3.

6.5 Limits of detection in filter weighing

Table 5 gives typical limits of detection for the methods described in the previous sections.

Table 5. Typical limits of detection for frequently used types of filters in filter weight determination according to Sections 6.4.1 to 6.4.3.

Type of filter (material, diameter (Ø) and, where applicable, pore size (PS))	Limit of detection		
	Laboratory blank filter (Method 1)	Calibration with environmental data (Method 2)	Blank filter during sampling (Method 3)
Cellulose nitrate, Ø 47 mm, PS 8.0 μm	0.15 mg	0.60 mg	
Cellulose nitrate, Ø 37 mm, PS 8.0 μm	0.15 mg	0.30 mg	
Cellulose nitrate, Ø 150 mm, PS 8.0 μm	1.50 mg	2.00 mg	
Cellulose nitrate, Ø 50 mm, PS 8.0 μm	0.15 mg		
Cellulose nitrate, Ø 50 mm, PS 3.0 μm	0.15 mg		
Cellulose nitrate, Ø 70 mm, PS 8.0 μm	0.15 mg	0.60 mg	
Quartz fibre filter S+S QF20, Ø 37 mm	0.10 mg		0.09 mg
Quartz fibre filter S+S QF20, Ø 47 mm	0.20 mg		0.14 mg
Quartz fibre filter Munktell MK360, Ø 70 mm	0.10 mg		
Glass fibre filter S+S GF6, Ø 25 mm			0.12 mg
Glass fibre filter MN 85/90BF, Ø 37 mm	0.30 mg	0.30 mg	0.30 mg
Glass fibre filter MN 85/90BF, Ø 47 mm	0.30 mg		
Glass fibre filter MN 85/90BF, Ø 70 mm		0.60 mg	0.20 mg
Glass fibre filter MN 85/90BF (S+S GF6), Ø 150 mm		3.00 mg	0.20 mg*
PU foam filters for the "A" dust fraction in the CIP10 sampler	1.00 mg		

* The filters used to calculate the limit of detection were treated according to the same working steps as the measurement filters, except they were not mounted in the sampling head.

6.6 Sources of errors

The influential factors listed below can lead to errors when dust masses are determined by weighing. Method development should investigate and minimise these effects by appropriate measures:

- Impurities resulting from filter capsules not providing a tight seal during transport.
- Loss of dust from the sample carrier as a result of filter cassettes not providing a tight seal during transport.
- Use of unsuitable containers for transport.
- Excessive heating or cooling of samples during transport.
- Adherence of dust inside the filter cassette.
- Change in filter weights as a result of fluctuations in humidity (absorption of humidity).
- Inadequate conditioning of the filters in the laboratory prior to weighing.

- Lacking or inappropriate checks of the environmental conditions in the laboratory.
- Static charges on filters or balance.
- Contamination of the balance by dust falling off the filters.
- Damage to filters on account of manual handling during the processing of sample carriers, weighing or sampling (filters pressed down too strongly during insertion of the filter cassettes into the sampling device).
- Use of filters from different batches in one measurement series.
- Dust pollution in the laboratory; e.g. conditioning uncovered filters and parallel processing of samples in the same laboratory.
- Use of filters beyond shelf-life.
- Installation of the balance in an unsuitable place (to be avoided: weighing tables that are not solid; transfer of vibrations to the balance e.g. because the weighing station lacks shock isolation; air draught or turbulence, direct sunlight, nearby heater/source of heat radiation, close proximity to doors or windows, proximity to strong electromagnetic fields; unstable power supply to the balance).
- Inadequate studies to determine the limit of detection (studies only carried out in the laboratory, not taking into account the effects of transport or handling of sample carriers when inserting them into the sampling device).

Manufacturing-related residual material from the confectioning process (punching out) may, in exceptional cases, be adhering to the filters. Such filters must not be used for sampling.
Specific problems are encountered in the determination of very hygroscopic dust. Weight changes based on this effect cannot be compensated by the weighing procedure. This may necessitate detailed studies.

7 Determination of metal-containing components of dust

In addition to the gravimetric determination of the dust weight on the collection medium, it is frequently necessary to carry out specific analyses for workplace-relevant hazardous substances contained in the dust.
Analysis for metals/metalloids predominantly resorts to methods (see Table 6) which require that the dust sample be brought into solution, meaning that the metals and metal compounds contained in the dust need to be extracted, dissolved or digested. The aim of this is to completely dissolve all relevant substances, i.e. all metals of toxicological significance.
As a rule, the elemental content (total concentration) of metals present in the sample is determined without regard to type of bonding or oxidation state. From the perspective of occupational medicine and toxicology, it is reasonable to distinguish between different compounds of a metal because the type and extent of harmful effects of metals depend considerably on their binding state and solution behaviour in the body. There may also be differences in effect of the same compound, depending on the route of exposure. In most instances, analytical methodology is unable to fulfil this demand for differentiation, or it can achieve differentiation only at considerable cost and effort.

The largely standardised digestion methods presented below have proved reliable in practice [30]. The conditions have been selected such that a maximum solubility effect can be assumed for most metals and metal compounds with respect to the dissolution processes in the body. It is important in this context to note that it is not always mandatory to bring the entire specimen into solution.

Generally, the digestion methodology should be selected to suit the sample to be examined. As a rule, however, it is difficult to implement this demand consistently because the material composition of the dusts is largely unknown. On-site research only provides an indication of the metals and metal compounds that can be expected to be present in higher concentrations in the dust. As a rule, there is little information on companion substances, which are most often present in higher concentrations than the substances to be determined.

Table 6. Frequently used analytical methods for workplace measurements of metals.

Method	Analytical technique
Atomic absorption spectroscopy (AAS)	Flame and graphite furnace methods (electrothermal atomic absorption spectrometry, ET-AAS) Hydride technique Cold vapour technique
Atomic emission spectrometry (AES)	Inductively-coupled plasma optical emission spectrometry (ICP-OES) Inductively-coupled plasma mass spectrometry (ICP-MS)
Atomic fluorescence spectrometry (AFS)	
X-ray fluorescence analysis (XRF)	Energy dispersive Wavelength dispersive Total reflection
Spectrophotometry (UV/VIS)	
Liquid chromatography (LC)	Ion chromatography (IC) High Performance Liquid Chromatography (HPLC)

Digestion methods employed should meet the following requirements:

- The substances to be determined (analytes, "substances of interest" according to [31]) must fully be brought into solution.
- Combined methods and multi-stage digestion should be avoided.
- All working steps should preferably be carried out in one reaction vessel.
- Contamination must be avoided.
- Losses of volatile compounds already present in the dust or formed during processing of the dusts must be prevented by appropriate measures (reflux condenser).
- Concentrations of contaminating salts in the digestion solution should be kept low so as to avoid interference during measurement (matrix effects).

The following influential factors need to be taken into consideration [32, 33]:

– Filter material (see Section 2.1.3)
– Particle size
– Particle size range is, as a rule, clearly defined by the measurement technique (inhalable and respirable fractions)
– Ratio of air sample volume to extraction volume
   (Dust solubility depends on the dust weight collected and the amount of reagent provided. Therefore the ratio of dust weight to extraction volume should be approximately constant.)
– Solvents and extractants
   (The amount depends on the type and weight of dust (or, alternatively, sampling volume) collected and on the type and size of the filters used.
   It is important when selecting solvents or extractants to ensure that they do not form sparingly soluble compounds with the dust components to be analysed.)
– Temperature and duration of treatment
   (In practice, relatively short treatment at higher temperatures, if necessary pressure or microwave-assisted, has proved reliable. This way, it is easier to eliminate possible interference from companion substances present in the dust.)

Filters should not be cut up because on the one hand, the homogeneity of dust loadings cannot be optically ascertained and, on the other, the relative limit of quantification would suffer (smaller amount of dust for analysis).

7.1 Preparation of dust samples for the determination of "soluble metal compounds"

This method of sample preparation is to precede the determination of metals and metal compounds, the limit values for which are linked with the term "soluble" (Table 7).
There is no clear or uniform definition of the term "soluble compounds". The "ad-hoc metals subgroup" within CEN/TC 137/WG 2 defines "soluble metal compounds" in terms of the specific extraction agents and conditions – as described by measuring procedures. The method described in the following is also mentioned in EN 13890 [31] (Annex A, Table 3, Sample preparation, Method 1).

Table 7. Limit values for soluble components of dust (examples).

Substance	Limit value ("E", or inhalable, fraction) [mg/m^3]	
	DFG MAK Values List [4]	TRGS 900 [7]
Barium compounds, soluble	0.5	0.5
Molybdenum compounds, soluble	Currently no MAK value	5.0
Niobium compounds, soluble	–	0.5
Thallium compounds, soluble	Currently no MAK value	0.1
Tungsten compounds, soluble	Currently no MAK value	1.0
Zirconium compounds, soluble	Currently no MAK value	–

The recommended extractant is diluted hydrochloric acid. If water is used as the extractant, it is possible that companion substances in the collected dust cause great variability in the physicochemical properties (e. g. pH or redox potential) of the resulting suspensions. This may lead to changes in the solution behaviour of identical metal compounds and, hence, to variation in the measurement results.

The use of buffer solutions or synthetic body fluids also cannot be recommended in daily practice due to manifold problems in the preparation of dust samples. Particularly negative effects on measurement reproducibility occur when these extraction agents are used with durations of treatment of up to 25 hours and treatment temperatures of only about 40 °C, as is often suggested.

The use of diluted hydrochloric acid helps to obviate many of these problems. Hydrochloric acid has no oxidative properties and most chlorides are readily soluble. Alternatively, silver determinations can be performed with diluted nitric acid (silver forms sparingly soluble compounds with hydrochloric acid) or concentrated hydrochloric acid (silver then goes into solution as a complex ion).

The working conditions for the determination of soluble compounds are summarised in Table 8.

Table 8. Working conditions for the determination for soluble compounds.

Parameter	Working conditions	
	For air sample volumes up to 600 litres and filter diameters up to 40 mm	For air sample volumes >600 litres up to 180 m^3 and filter diameters up to 150 mm
Extractant	0.1 M hydrochloric acid	
Amount of reagent	20 mL	80 mL
Pressure	Normal pressure	Normal pressure
Temperature	Boiling temperature	Boiling temperature
Duration of treatment	2 hours	2 hours
Digestion vessel	Graduated digestion cylinder with standard ground glass socket joint (NS), equipped with a condenser and boiling rods	
	Volume: 25 mL Graduation: 0.2 mL divisions Ground joint: NS 19/26	Volume: 100 mL Graduation: 1.0 mL divisions Ground joint: NS 29/32

Procedure

The loaded filter is placed in the digestion vessel and the extractant is then added. The amount added depends on the sampling conditions (see Table 8). The digestion vessels are equipped with boiling rods and air condensers (length approx. 40 cm) and heated under reflux for two hours in a thermostat-controlled aluminium heating block provided with the necessary holes (block temperature approx. 125 °C). A volume reading of the solution is taken after cooling and sedimentation of the residue, and the supernatant is used for analysis. If sedimentation is incomplete, an aliquot of the solution is filtered before analysis.

The above-described procedure can be modified to suit the equipment used.

7.2 Preparation of dust samples for the determination of "total metal content"

This method of sample preparation is to precede the determination of metals and metal compounds, the limit values for which refer to the total metal content of the dust.

A mixture of hydrochloric acid and nitric acid is used (see Table 9 below). Nitric acid is employed because of its properties as an excellent solvent and powerful oxidant, while hydrochloric acid acts as a solubiliser. The high concentration of hydrochloric acid present after digestion yields solutions of relatively high viscosity, which can cause considerable interference in the analytical determination (e. g. by AAS). It has therefore proved effective to introduce further dilution steps prior to analysis.

The working conditions for the determination of total metal content in dust are described in Table 9 below.

Table 9. Working conditions for the determination for total metal content.

Parameter	Working conditions	
	For air sample volumes up to 600 litres and filter diameters up to 40 mm	For air sample volumes > 600 litres up to 180 m^3 and filter diameters up to 150 mm
Extractant	2 parts by volume of nitric acid ($\geq 65\%$) and 1 part by volume of hydrochloric acid (25%)	
Amount of reagent	10 mL	40 mL
Dilution after sampling*	10 mL (distilled H$_2$O)	40 mL (distilled H$_2$O)
Pressure	Normal pressure	Normal pressure
Temperature	Boiling temperature	Boiling temperature
Duration of treatment	2 hours	2 hours
Digestion vessel	Graduated digestion cylinder with standard ground glass socket joint (NS), equipped with a condenser and boiling rods	
	Volume: 25 mL Graduation: 0.2 mL divisions Ground joint: NS 19/26	Volume: 100 mL Graduation: 1.0 mL divisions Ground joint: NS 29/32

* Interference-free measurement requires a further dilution step. As a rule, a 1 : 4 dilution (1 part of digestion solution + 3 parts of water) is sufficient.

Procedure

The whole loaded filter is placed in the digestion vessel and the recommended amount of extraction agent is then added. The amount added depends on the sampling conditions (see Table 9). The digestion vessels are equipped with boiling rods and air condensers (length approx. 40 cm) and heated under reflux for two hours in a thermostat-controlled aluminium heating block provided with the necessary holes (block temperature approx. 125 °C). A volume reading is taken after cooling and careful dilution with water via the condenser. A further dilution step is usually necessary prior to analysis. This is achieved by diluting an aliquot of digestion solution with three parts of water. The solution thus obtained can then be analysed.

The above-described procedure can be modified to suit the equipment used. It is important when using open systems to bear in mind that some metals are capable of forming highly volatile compounds, and that this can result in low recoveries.

7.3 Notes on the methods described and speciation analysis

The methods described here are recommendatory in nature. Experience shows, however, that the suggested methods yield concordant results for the majority of problems. The procedure has also proved itself with respect to the time and effort needed for sample preparation. Reproducibility of the measured values benefits from sample preparation according to largely standardised procedures. Moreover, round robin studies [34] have demonstrated that compared with considerably more elaborate procedures there is no significant difference in the accuracy of analytical results even if the deposited dust is not completely dissolved (see Appendix 4). None the less, solutions that are free of optically detectable residues are always recommended.

The extent to which further analysis of a residue is mandatory for the evaluation of the workplace exposure situation needs to be decided on a case-by-case basis. With specific samples or analytes, expanded or different digestion methods can yield better results, particularly when additional information on the sample is available.

For instance, acids or mixtures of acids (hydrochloric acid, nitric acid, hydrofluoric acid, perchloric acid) can be varied with respect to quantity, concentration and mixing ratio. Suitable methods for sparingly soluble alloying constituents could include (microwave-assisted) pressure digestions in closed systems and at higher temperatures. Fusion with subsequent dissolution of the melt cake is rarely necessary with airborne dusts.

The documentation "Empfohlene Analysenverfahren für Arbeitsplatzmessungen" (Recommended analytical procedures for workplace measurements; GA 13) [9] by the "Bundesanstalt für Arbeitsschutz und Arbeitsmedizin" (BAuA; Federal Institute for Occupational Safety and Health) provides valuable information on methods for the quantitative analysis of metal-containing airborne dusts.

The digestion and measurement methods described above such as AAS or ICP-OES allow the determination of metals and metal compounds in dust in terms of elemental content only. This is sufficient for most types of problems, since many limit values refer to the concentration of metal (calculated on the basis of the element). Analytical methodology often is not able to provide differentiation according to compounds or oxidation states (speciation analysis), as often demanded for toxicological reasons. It is recommended in such cases to adopt a worst-case approach, comparing the elemental concentration in air (after conversion using the appropriate stoichiometric factor) with the limit values used in occupational hygiene.

Where available, however, accepted, specific methods should be used, e.g. for Cr (VI) or Sb (III, V) [35, 36].

In recent years, there has been progress in the area of speciation analysis – i.e. the determination of the oxidation state of an element and the bonds it forms with organic or

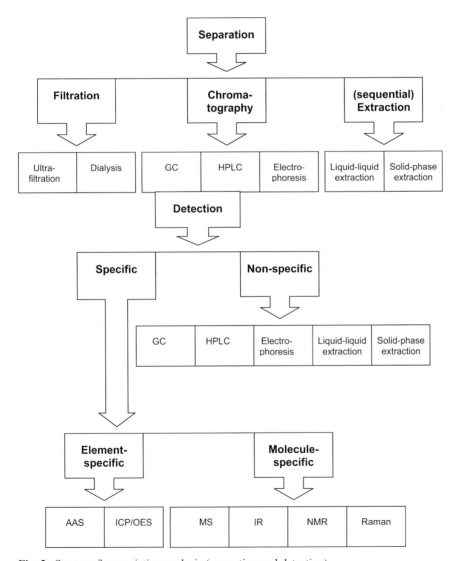

Fig. 2. Strategy for speciation analysis (separation and detection).

inorganic binding partners or the characterisation of different analyte molecules in a matrix. However, the questions addressed were usually very specific and not primarily studied with a view to wider applicability.

As is the case even in elemental analysis as it is generally practiced, sample collection and matrix-appropriate sample preparation are the most important part of the measurement procedure. It is especially important in the case of speciation analysis to ensure that no changes (e.g. in pH or redox conditions) occur in the sample during sampling or sample preparation, which could lead to incorrect results.

In speciation analysis [37], sample preparation is followed first by analytical separation of the individual species or groups of species, which then in turn is followed by an element- or molecule-specific method of species identification (see Figure 2). So, what is essential in speciation analysis is that it combines methods of separation and analytical methods, something that is often achieved only by using complex coupled techniques. A new way of carrying out molecule-specific searches and detecting hazardous substances is the use of biotechnological measurement systems (biosensors).

7.4 Quality control

The complete measuring procedure should be regularly subjected to quality control. This can be achieved in the form of internal quality assurance measures or by participating in round robin studies in which dust-loaded filters or dusts are analysed. Internal quality assurance can be achieved by repeated analysis of certified reference materials (dusts). This allows checks on the precision and accuracy of the method. The effect of sampling on the analytical result, however, is not then taken into account.

8 References

[1] *Barig A. and Blome H.* (1999) Allgemeiner Staubgrenzwert. Teil 2: Arbeitsplatzexposition, Aspekte der praktischen Umsetzung. Gefahrstoffe – Reinh. Luft 59, Nr. 11/12: 409–417

[2] *Europäisches Komitee für Normung* (CEN) (1993) DIN EN 481 – Arbeitsplatzatmosphäre – Festlegung der Teilchengrößenverteilung zur Messung luftgetragener Partikel. Brüssel. Beuth-Verlag, Berlin. (Workplace atmospheres – Size fraction definitions for measurement of airborne particles)

[3] *Proceedings of the Pneumoconiosis Conference* (1959) Recommendations of the Johannesburg Pneumoconiosis Conference 1959. Johannesburg, p. 619.

[4] *Deutsche Forschungsgemeinschaft* (2003) List of MAK und BAT Values 2003. Commission for the Investigation of Health Hazards of Chemical Compounds in the Work Area. Report No. 39. WILEY-VCH, Weinheim

[5] *Berufsgenossenschaftliches Institut für Arbeitsschutz* (BIA) (Ed.) (2003) Messung von Gefahrstoffen – BIA-Arbeitsmappe. Expositionsermittlung bei chemischen und biologischen Einwirkungen. 0412: Allgemeine Staubgrenzwerte, Erich Schmidt Verlag, Bielefeld, Loseblatt-Ausgabe

[6] *Bundesministerium für Arbeit und Sozialordnung* (1997) TRGS 402: Ermittlung und Beurteilung der Konzentration gefährlicher Stoffe in der Luft in Arbeitsbereichen, BArbBl. 11/1997, S. 27–33

[7] *Bundesministerium für Wirtschaft und Arbeit* (2004) TRGS 900: Grenzwerte in der Luft am Arbeitsplatz – Luftgrenzwerte, BArbBl. 10/2000, S. 34–63, zuletzt geändert BArbBl. 5/2004, S. 55–65

[8] *Bundesministerium für Arbeit und Sozialordnung* (2001) TRGS 901: Begründungen und Erläuterungen zu Grenzwerten in der Luft am Arbeitsplatz, Nr. 96: Allgemeiner Staubgrenzwert, BArbBl. 9/2001, S. 89

[9] *Auffarth J., Hebisch R. and Poppek U.* (2003) Empfohlene Analysenverfahren für Arbeitsplatzmessungen. Schriftenreihe der Bundesanstalt für Arbeitsschutz und Arbeitsmedizin – Gefährliche Arbeitsstoffe (GA 13), Wirtschaftsverlag NW, Bremerhaven

[10] *Greim H.* (Ed.) (1996) Analytische Methoden zur Prüfung gesundheitsschädlicher Arbeitsstoffe, Luftanalysen, Band 1, Dieselmotoremissionen Meth.-Nr. 1 und 2, Wiley-VCH, Weinheim

[11] *Greim H.* (Ed.) (1996) Analytische Methoden zur Prüfung gesundheitsschädlicher Arbeitsstoffe, Luftanalysen, Band 1, Quarz Meth.-Nr. 1, Wiley-VCH, Weinheim

[12] *Europäisches Komitee für Normung* (CEN) (1994) DIN EN 482 – Arbeitsplatzatmosphäre – Allgemeine Anforderungen an Verfahren für Messung von chemischen Arbeitsstoffen. Brüssel. Beuth-Verlag, Berlin. (Workplace atmospheres – General requirements for the performance of procedures for the measurement of chemical agents)

[13] *Europäisches Komitee für Normung* (CEN) (2002) DIN EN 13205 – Arbeitsplatzatmosphäre – Bewertung der Leistungsfähigkeit von Geräten für die Messung der Konzentration luftgetragener Partikel. Brüssel. Beuth-Verlag, Berlin. (Workplace atmospheres – Assessment of performance of instruments for measurement of airborne particle concentrations)

[14] *Europäisches Komitee für Normung* (CEN) (1997) DIN EN 1232 – Arbeitsplatzatmosphäre – Pumpen für die personenbezogene Probenahme von chemischen Stoffen: Anforderungen und Prüfverfahren. Brüssel. Beuth-Verlag, Berlin. (Workplace atmospheres – Pumps for personal sampling of chemical agents – Requirements and test methods)

[15] *Europäisches Komitee für Normung* (CEN) (1999) DIN EN 12919 – Arbeitsplatzatmosphäre – Pumpen für die Probenahme von chemischen Stoffen mit einem Volumendurchfluss über 5 l/min: Anforderungen und Prüfverfahren. Brüssel. Beuth-Verlag, Berlin. (Workplace atmospheres – Pumps for the sampling of chemical agents with a volume flow rate of over 5 l/min – Requirements and test methods)

[16] *Europäisches Komitee für Normung* (CEN) (2000) DIN EN 50014 – Elektrische Betriebsmittel für explosionsgefährdete Bereiche – Allgemeine Bestimmungen. Brüssel. Beuth-Verlag, Berlin. (Electrical apparatus for potentially explosive atmospheres – General requirements)

[17] *Greim H.* (Ed.) (2000 and 2003) Analytische Methoden zur Prüfung gesundheitsschädlicher Arbeitsstoffe, Luftanalysen, Band 1, Lackaerosole Meth.-Nr. 1–3, Wiley-VCH, Weinheim

[18] *Europäisches Komitee für Normung* (CEN) (2000) DIN EN 143 – Atemschutzgeräte – Partikelfilter: Anforderungen, Prüfung, Kennzeichnung. Brüssel. Beuth-Verlag, Berlin. (Respiratory protective devices – Particle filters – Requirements, testing, marking)

[19] *Verein Deutscher Ingenieure* (1980) VDI-Richtlinie 2265 – Feststellen der Staubsituation am Arbeitsplatz zur Gewerbehygienischen Beurteilung. Beuth-Verlag, Berlin

[20] *Möhlmann C., Aitken R. J., Kenny L. C., Görner C., VuDuc T. and Zambelli G.* (2002) Size-selective personal air sampling: a new approach using porous foams. Ann. Occup. Hyg. 46, Suppl. 1: 386–389

[21] *Berufsgenossenschaftliches Institut für Arbeitsschutz* (BIA) (2002) Porous foam aerosol sampling (PoFaS). Report on the EU project "Size selective personal air sampling using porous plastic foams. Co-ordinator: BIA, St. Augustin

[22] *Möhlmann C., Aitken R. J., Kenny L. C., Görner C., VuDuc T. and Zambelli G.* (2003) Größenselektive personenbezogene Staubprobenahme: Verwendung offenporiger Schäume. Gefahrstoffe – Reinh. Luft 63, Nr. 10: 413–416.

[23] *Europäisches Komitee für Normung* (CEN) (1995) DIN EN 689 – Arbeitsplatzatmosphäre – Anleitung zur Ermittlung der inhalativen Exposition gegenüber chemischen Stoffen zum Vergleich mit Grenzwerten und Messstrategie. Brüssel. Beuth-Verlag, Berlin. (Workplace atmospheres – Guidance for the assessment of exposure by inhalation to chemical agents for comparison with limit values and measurement strategy)

[24] *Bundesministerium für Arbeit und Sozialordnung* (1989) TRGS 403: Bewertung von Stoffgemischen in der Luft am Arbeitsplatz, BArbBl. 10/1989, S. 71–72

[25] *Länderausschuss für Arbeitsschutz und Sicherheitstechnik* (LASI) (1995) LASI-Veröffentlichungen LV 2: Richtlinien für die Akkreditierung von außerbetrieblichen Messstellen zum Vollzug des Gefahrstoffrechts gemäß § 18 Abs. 2 Gefahrstoffverordnung, Wiesbaden

[26] *Europäisches Komitee für Normung* (CEN) (2000) DIN EN ISO/IEC 17025 – Allgemeine Anforderungen an die Kompetenz von Prüf- und Kalibrierlaboratorien. Brüssel. Beuth-Verlag, Berlin. (General requirements for the competence of testing and calibration laboratories)

[27] *Bundesministerium für Arbeit und Sozialordnung* (1975) Verordnung über Arbeitsstätten. BGBl I 1975, S. 729, zuletzt geändert durch Art. 281 V v. 25.11.2003, BGBl I, S. 2304

[28] *Deutsches Institut für Normung* (2004) DIN ISO 15767 – Arbeitsplatzatmosphäre – Kontrolle und Charakterisierung der Fehler beim Wägen gesammelter Aerosole. Beuth-Verlag, Berlin.

[29] *Berufsgenossenschaftliches Institut für Arbeitsschutz* (BIA) (Ed.) (2003) Messung von Gefahrstoffen – BIA-Arbeitsmappe. Expositionsermittlung bei chemischen und biologischen Einwirkungen. 6068: Messverfahren für Gefahrstoffe – alveolengängige Fraktion, Erich Schmidt Verlag, Bielefeld, Loseblatt-Ausgabe

[30] *Hahn, J. U.* (2000) Aufarbeitungsverfahren zur analytischen Bestimmung löslicher Metallverbindungen. Ein pragmatischer Vorschlag. Gefahrstoffe – Reinh. Luft 60, Nr. 6: 241–243

[31] *Europäisches Komitee für Normung* (CEN) (2003) DIN EN 13890 – Arbeitsplatzatmosphäre – Verfahren zur quantitativen Bestimmung von Metallen und Metalloiden in luftgetragenen Partikeln – Anforderungen und Prüfverfahren. Brüssel. Beuth-Verlag, Berlin. (Workplace atmospheres – Procedures for measuring metals and metalloids in airborne particles – Requirements and test methods)

[32] *Berufsgenossenschaftliches Institut für Arbeitsschutz* (BIA) (Ed.) (2003) Messung von Gefahrstoffen – BIA-Arbeitsmappe. Expositionsermittlung bei chemischen und biologischen Einwirkungen. 6015: Aufschlussverfahren zur Analytik metallhaltiger Stäube, Erich Schmidt Verlag, Bielefeld, Loseblatt-Ausgabe

[33] *Bock R.* (2001) Handbuch der analytisch-chemischen Aufschlussmethoden, Wiley-VCH, Weinheim

[34] *Breuer D.* (1997) Ringversuche für innerbetriebliche Messstellen Eine kurze Ergebnisübersicht der Ringversuche des Jahres 1996, BIA-Mitteilungen zur Arbeitsbereichsüberwachung in Betrieben (BAB-Info). Gefahrstoffe – Reinh. Luft 57, Nr. 7/8: 303–304

[35] *Kettrup A.* (Ed.) (1999) Analyses of Hazardous Substances in Air, Vol. 4, Hexavalent chromium, Wiley-VCH, Weinheim

[36] *Kettrup A.* (Ed.) (2003) Analyses of Hazardous Substances in Air, Vol. 7, Antimony trioxide, Wiley-VCH, Weinheim

[37] *Cornelis R., Caruso J., Crews H. and Heumann K.* (Eds.) (2003) Handbook of Elemental Speciation – Techniques and Methodology, John Wiley & Sons, Chichester

9 Appendices

Appendix 1: Overview of pumps for personal sampling

Pump	Working range mL/min	Supplier
Gilian PP5-Ex/HFS 513 AUP	750–5000	1, 2, 3
GilAir-3 Ex	500–3000	1, 2, 3
GilAir-5 Ex	750–5000	1, 2, 3
Gilian 3500	700–3500	1, 2, 3
ALPHA 1 (no longer in production)	5–5000	Service only
SG 10	1000–10000	1, 2, 3
SKC 224 EX	5–5000	1, 4
SKC 224 XR	5–4000	1, 4
SKC-EXECUTIVE	5–3250	1, 4
SKC Sidekick	5–3000	1, 4
GSA 5002 ex (no longer in production)	500–5000	Service only
S 2500 ex (no longer in production)	300–2500	Service only
P 4000 (no longer in production)	20–4000	Service only
ESCORT ELF	500–3000	2
GSA SG 4500	10–4000	3
Buck-Genie VSS-5	800–5000	5

Suppliers:

1 = DEHA Haan + Wittmer GmbH
    Postfach 11 43
    D-7259 Friolzheim

2 = LMT Leschke Messtechnik GmbH
    Bergstraße 168
    D-15230 Frankfurt/Oder

3 = GSM Gesellschaft für Schadstoffmesstechnik
    Gut Vellbrüggen
    D-41469 Neuss

4 = ANALYT/MTC-GmbH
    Postfach 13 21
    D-79379 Mülheim

5 = Ravebo Supply B.V.
    't Woud 2
    NL-3230 AG Brielle

Appendix 2: Overview of sampling heads for collecting dust samples in workplace measurements

Sampling head	Flow rate m³/h	Separation principle	Dust fraction collected
VC 25F	22.5	Impaction	"A" dust
VC 25I	22.5	Impaction with an upstream impactor for droplet aerosols	(respirable dust)
PM 4F	4.0	Cyclone	
MPG II/MPG III	2.8	Sedimentation pre-separator (horizontal elutriator)	
Casella cyclone	0.12	Cyclone	
FSP-BIA	0.12	Cyclone	
FSP 10	0.60	Cyclone	
VC 25 G	22.5	Ring-gap inlet	"E" dust
PM 4G	4.0	Inlets	(inhalable dust)
STASA	0.15	Conical inlet cylinder	
GSP-BIA	0.21	Suction cone	
GSP 10	0.60	Suction cone	
PGP-EA	0.21	Suction cone	"A" dust
PGP-ETA	0.21	Suction cone	+
Respicon	0.19	Ring-gap inlet	"E" dust
TBF 50	3.0	Cyclone	

Appendix 3: Examples of filter materials and collection media for dust sampling

Filter materials	Type, supplier
Cellulose nitrate filters	Sartorius Type 11301, Sartorius AG, D-37075 Göttingen AE99, Schleicher & Schuell GmbH, D-37586 Dassel
Cellulose acetate	Sartorius Type 11104, Sartorius AG, D-37075 Göttingen
Glass fibre filters	MN 85/90 BF, Macherey-Nagel, D-52355 Düren GF 6, Schleicher & Schuell GmbH, D-37586 Dassel
Quartz fibre filters	QF 20, Schleicher & Schuell GmbH, D-37586 Dassel Ederol T293, Binzer & Munktell, D-35088 Battenberg Munktell Quartz Microfibre MK 360, Binzer & Munktell, D-35088 Battenberg
Tissue quartz	Pall QAO 2500, Pall GmbH, D-63303 Dreieich
Teflon filters	Zefluor, Pall GmbH, D-63303 Dreieich
Inpactor/selector/ rotating sponge	For dust collection system CIP 10, Arelco – Dr. Ing. Wazau, D-10589 Berlin

Appendix 4: Effect of the digestion method on the precision and accuracy of analytical
             results

Round robin studies can be used to perform analytical quality assurance. The use of
dusts or dust-loaded filters as samples offers the advantage that the digestion method
can also be included in the assessment.

The extent to which digestion affects the result was demonstrated by two round robin
tests conducted by BIA. Tests were carried out as follows, each with approx. 0.5 g of
the same certified dust:

– In the first round robin test (59 participants), there was a free choice as to the
  method of dust digestion. Analyses were performed for arsenic, lead and nickel.
– In the second round robin test (51 participants), use of the standardised digestion
  method described in Section 7.2 was mandatory. Analyses were performed for chromium, copper and nickel.

All participating laboratories were required to provide exact descriptions of the digestion conditions they used, thus enabling assessment with regard to digestion. The methods used in the first round robin study were numerous and often quite different, as
shown in the following summary.

Effect parameter	Number of methods	Predominant method
Digestion media (acids used)	18 different mixtures	Mixtures of HCl and HNO_3 (64% of participating laboratories)
Amount of reagent (per 100 mg of dust)	22 different statements	25 mL (49% of participating laboratories)
Duration of treatment	12 different statements	2–3 hours (54% of participating laboratories)
Pressure digestion	12 different pressures	Normal pressure (73% of participating laboratories)

It was found that practically every laboratory used its own digestion method. In some
cases, elaborate and complicated methods yielded deviations from the nominal value
that were considerable.

In the second round robin study, approx. 80% of the participating laboratories complied
with the prescribed digestion method. Since the same dust was sent out as in the first
round robin study, the results for nickel are directly comparable. They are summarised
below.

Round robin study No.	$C_{nominal}$ µg/g	C µg/g	S_{rel} %	N	No. of outliers
1	100	93	19.3	59	15
2	100	89	12.2	51	10

$C_{nominal}$ = certified value, C = mean value of the laboratory results, S_{rel} = standard deviation, N = number of laboratories

Comparison of the results from the two round robin studies clearly illustrates that better results are obtained with the recommended method. Harmonisation of the digestion method leads to enhanced reproducibility of results without significant losses in accuracy.

Acknowledgement

The authors wish to thank the members of the DFG working group on air analyses (DFG-Arbeitskreis "Luftanalysen") and the DFG working group on limit values for dusts (DFG-Arbeitsgruppe "Festlegung von Grenzwerten für Stäube") for their interesting and helpful contributions to the discussions.

Authors: *Ralph Hebisch, Hajo-Hennig Fricke, Jens-Uwe Hahn, Majlinda Lahaniatis, Claus-Peter Maschmeier, Markus Mattenklott*

Analytical Methods

Ammonia

Method number   2

Application   Air analysis

Analytical principle   Ion chromatography

Completed in   May 2003

Summary

The present analytical method permits the determination of ammonia within a concentration range of 0.04–2 times the threshold limit value proposed by the Deutsche Forschungsgemeinschaft. Sampling is conducted with a suitable pump by drawing ambient air through a combination sample carrier consisting of a Teflon filter and a glass tube packed with finely dispersed carbon (carbon bead) impregnated with acid [1]. During this procedure, particulate ammonium compounds are retained by the filter whilst gaseous ammonia is retained by the carbon bead. After sampling, the ammonia-loaded sampling and backup sections from the adsorption tube are covered with elution solution. After desorption in an ultrasonic bath, the elution solutions are analysed by ion chromatography. The component is quantified by means of a conductivity detector.

Characteristics of the method

Accuracy:

Table 1. Standard deviation s_{rel} and mean variation u, $n = 6$ determinations.

Concentration	Standard deviation s_{rel}	Mean variation u
mg/m^3	%	%
0.51	1.31	3.3
1.43	0.80	2.1
14.4	2.85	7.3
26.4	1.52	3.9

The MAK-Collection Part III: Air Monitoring Methods, Vol. 9. DFG, Deutsche Forschungsgemeinschaft
Copyright © 2005 WILEY-VCH Verlag GmbH & Co. KGaA, Weinheim
ISBN: 3-527-31138-6

Limit of quantification: 20 ng absolute 0.09 mg/m^3 for an air sample volume of
                            40 litres

Recovery: 105%

Sampling recommendation: Sampling time: 2 hours
                            Air sample volume: 40 litres

Ammonia [CAS No. 7664-41-7]

Ammonia (NH_3) is a colourless toxic and lachrymatory gas with a pungent odour (molar mass 17.03 g/mol, melting point -78 °C, boiling point -33 °C) which is readily soluble in water, alcohol, benzene, acetone and chloroform. Ammonia vapours cause irritation even at low concentration levels and are corrosive to the mucous membranes of the eyes and respiratory tract at higher levels. Because of the intensely pungent odour, however, ammonia intoxications are very rare.

The Deutsche Forschungsgemeinschaft proposes a threshold limit value of 14 mg/m^3 (2.0 mL/m^3) in air [2] and placement in peak limitation category I(2). According to the TRGS 900, the currently valid threshold limit value for air is 35 mg/m^3 (50 mL/m^3). The peak limitation for ammonia has been assigned an excursion factor of $=1=$ [3].

Ammonia is a basic product of the chemical industry and serves as a starting material for chemical synthesis in the manufacture of, e. g., urea, sulfonamides, synthetic fibres, sodium cyanide, hydrogen cyanide and nitriles, sodium carbonate, aminoplastics, nitric acid and nitrates. Ammonium salts are used inter alia as fertilisers.

Liquid ammonia is used e. g. in refrigerating machines and in textile finishing. In the gaseous state, the compound is used e. g. as a protective gas and in flue-gas desulfurisation. Ammonia water is used for cleaning purposes and as a mordant, as a neutralising agent and for adjusting the pH of solutions to specific values.

The present method was checked for plausibility by experts from the DFG working group on air analyses.

Authors: *D. Breuer, B. Heinrich*
Examiner: *R. Hebisch*

Ammonia

Method number   2

Application   Air analysis

Analytical principle   Ion chromatography

Completed in   May 2003

Contents

1 General principles

The present analytical method permits the determination of ammonia within a concentration range of 0.04–2 times the threshold limit value proposed by the Deutsche Forschungsgemeinschaft. Sampling is conducted with a suitable pump by drawing ambient air through a combination sample carrier consisting of a Teflon filter and a glass tube packed with finely dispersed carbon (carbon bead) impregnated with acid [1]. During this procedure, particulate ammonium compounds are retained by the filter whilst gaseous ammonia is retained by the carbon bead. After sampling, the ammonia-loaded sampling and backup sections from the adsorption tube are covered with elution solution. After desorption in an ultrasonic bath, the elution solutions are analysed by ion chromatography. The component is quantified by means of a conductivity detector.

2 Equipment, chemicals and solutions

2.1 Equipment

Pump for personal air sampling, flow rate 20 L/h
Gasmeter
Combination sample carrier (see Figure 1) consisting of:
– Filter holder, diameter 37 mm, e. g. Millipore: aerosol monitor
– Teflon filter, diameter 37 mm, pore size 0.45 µm, e. g. from Macherey-Nagel, Düren
– Carbon bead tube, impregnated with sulfuric acid, e. g. Orbo 77, Supelco, Sigma-Aldrich Chemie GmbH, Taufkirchen
Ion chromatograph with autosampler, column oven, suppressor and conductivity detector
Data system
25, 50 and 1000 mL Polypropylene volumetric flasks
Adjustable piston pipettes 10–10 000 µL
20 mL Screw-cap polyethylene containers
2500 mL Polypropylene storage bottle for storing the eluent
Autosampler vials made of polypropylene with screw caps and silicone/PTFE septa
Disposable filters for the filtration of aqueous samples, diameter 25 mm, pore size 0.45 µm, suitable for ion chromatography
Disposable syringes, 2 mL, with disposable needles (0.9 × 40 mm)
Ultrasonic bath
Ultrapure water system

2.2 Chemicals

Ammonium standard solution: w (NH$_4^+$) = 1000 µg/mL, e. g. from Merck,
                                                   Darmstadt
Ammonia solution: w (NH$_3$) = 25 %, e. g. from Merck, Darmstadt
Sulfuric acid: c (H$_2$SO$_4$) = 2.5 mol/L, e. g. Combi Titrisol,
                                                   from Merck

Dilutions of the 25 % (w/w) ammonia solution were used in all experiments carried out using the dynamic test atmosphere generator during the development of the method.

2.3 Solutions

Eluent: c (H$_2$SO$_4$) = 0.005 mol/L

A volume of 2 mL sulfuric acid is pipetted into a 1-litre volumetric flask and the flask is filled to the mark with ultrapure water (conductivity >18.2 M$\Omega \times$ cm).

2.4 Calibration standards

Stock solution: w (NH$_4^+$) = 50 µg/mL

A volume of 1250 µL ammonium standard solution is pipetted into a 25 mL volumetric flask and the flask is filled to the mark with eluent. The stock solution produced in this way has an ammonium concentration of 50 µg/mL and can be stored in the refrigerator at +4 °C for at least four weeks.

The volumes of the combined standard solution given in Table 2 are pipetted into 25 mL polypropylene volumetric flasks; these are filled to the mark with eluent, stoppered and shaken. The calibration solutions are transferred to autosampler vials and analysed. The calibration solutions are stable at room temperature for at least two weeks.

Table 2. Pipetting scheme for calibration standards for a concentration range of 0.6–6 µg/mL.

Calibration solution No.	Volume of the stock solution µL	Ammonium concentration µg/mL
1	300	0.6
2	600	1.2
3	900	1.8
4	1200	2.4
5	1500	3.0
6	1800	3.6
7	2100	4.2
8	2400	4.8
9	2700	5.4
10	3000	6.0

3 Sample collection and preparation

A pump equipped with a flow regulator is used to draw, at a flow rate of 0.333 L/min, the sample air through the combination sample carrier consisting of a Teflon filter and a glass tube packed with carbon bead.

The carbon bead adsorption tube is sealed with the plastic caps provided. The Teflon filter loaded with particulate ammonium compounds is discarded.

The sampling and backup sections from the carbon bead adsorption tube are placed in separate screw-cap containers and covered with 5 mL elution solution. The containers are sealed, treated in the ultrasonic bath for 15 minutes and allowed to cool to room temperature for about 30 minutes.

Liquid from the supernatant solution of the prepared sample is drawn into a disposable syringe and filtered through a membrane filter into an autosampler vial.

4 Operating conditions for chromatography

Apparatus:	Ion chromatography system, from Dionex GmbH, D-65510 Idstein, consisting of:
	GP 40 gradient pump
	ED 40 electrochemical detector
	AS 3500 autosampler
	Peaknet 5.1 chromatography software
Precolumn:	IonPac CG14, from Dionex
Length:	50 mm
Internal diameter:	4 mm
Column:	IonPac CS14, from Dionex
Length:	250 mm
Suppression:	CSRS-II cation self-regenerating suppressor, 4 mm, from Dionex
Column temperature:	40 °C
Mobile phase:	Sulfuric acid c (H$_2$SO$_4$) = 0.005 mol/L
Flow rate:	1.0 mL/min
Injection volume:	25 µL
System pressure:	approx. 10^5 hPa

Figure 2 shows an example of a chromatogram obtained under the conditions given above.

5 Analytical determination

A 25 µL aliquot of sample solution is injected via an autosampler and analysed under the conditions given above. If the concentrations determined are not within the calibration range, the samples must be diluted appropriately and analysed again.

6 Calibration

The calibration solutions described in Section 2.4 are used to construct a calibration curve.
Volumes of 25 µL of each of the calibration solutions are injected and analysed as for the sample solutions. The peak areas obtained are plotted against the corresponding concentrations. The calibration function is not linear in the range studied but rather represents a second-degree function. Figure 3 shows an example of an ammonium calibration curve.
Control samples should be analysed each working day to verify the calibration curve. Calibration must be repeated if the analytical conditions change or quality control shows this to be necessary.

7 Calculation of the analytical result

The concentration of ammonia in the workplace air is calculated using the ammonia concentration in the sample solution as calculated by the data system. The data system uses the second-degree calibration function calculated for the calibration curve for this purpose.

$$F = c \cdot w(NH_4^+) + b \cdot w(NH_4^+) + a \tag{1}$$

Based on the concentration in the sample solution, the concentration of ammonia in the workplace air is calculated, taking into account dilution steps, the blank value for the batch of carbon bead used and the air sample volume.
The following equations apply for calculation of the concentration of ammonia in the workplace air:

$$\rho = \frac{w(NH_4^+) \cdot 5 \cdot 0.9441}{V_{air} \cdot 1000} \cdot \frac{273 + t_g}{273 + t_a} \tag{2}$$

At 20 °C and 1013 hPa:

$$\rho_0 = \rho \frac{273 + t_g}{293} \cdot \frac{1013 hPa}{p_a} \tag{3}$$

The corresponding concentration by volume σ – independently of the state variables pressure and temperature – is given by:

$$\sigma = \rho_0 \frac{V_M}{M} \tag{4}$$

$$\rho = \rho \frac{273 + t_g}{p_a} \cdot \frac{1013 \text{ hPa}}{293} \cdot \frac{V_M}{M} \tag{5}$$

For ammonia at $t_a = 20\,^{\circ}\text{C}$ and $p_a = 1013$ hPa:

$$\sigma(\text{NH}_3) = \rho \cdot 1.316 \ \frac{\text{mL}}{\text{mg}} \tag{6}$$

where:

$w(\text{NH}_4^+)$	is the ammonium concentration in μg/mL
F	is the peak area
ρ	is the concentration by weight in mg/m^3 in the ambient air at t_a and p_a
ρ_o	is the concentration by weight in mg/m^3 in the ambient air at 20 °C and 1013 hPa
a	is the intercept of the calibration curve
b	is a constant in mL/μg
c	is a constant in (mL/μg)^2
5	is the conversion factor for the elution volume of the measured sample in mL
0.9441	is the stoichiometric conversion factor for ammonium to ammonia
1000	is the conversion factor for μg to mg
V_{air}	is the air sample volume in m^3
t_g	is the temperature in the gasmeter in °C
t_a	is the temperature of the ambient air in °C
p_a	is the atmospheric pressure of the ambient air in hPa
σ	is the ammonia concentration in the ambient air in mL/m^3
V_M	is the molar volume of ammonia in L/mol
M	is the molar mass of ammonia in g/mol

Where appropriate, the recovery must be taken into account when calculating the analytical results.

8 Reliability of the method

The characteristics of the method were determined according to the standard DIN EN 482 [4].

8.1 Precision

Precision was determined by vaporising, in a dynamic test atmosphere generator at a relative humidity of 40%, four solutions in the concentration range of approximately 0.04–2 times the MAK value for ammonia. Six sample carriers per concentration were loaded, processed and analysed under the sampling conditions described (Table 3).

Table 3. Relative standard deviation s_{rel} and mean variation u.

Concentration	Standard deviation	Mean variation
	s_{rel}	u
mg/m^3	%	%
0.51	1.31	3.3
1.43	0.80	2.1
14.4	2.85	7.3
26.4	1.52	3.9

8.2 Recovery

The recovery of ammonia was determined using the results for the determination of precision in the minimum measurement range. Based on the individual recoveries, mean recovery was calculated to be 105% ($\bar{\eta} = 1.05$).
The effect of higher humidity on ammonia sampling was also ascertained using the results for the determination of precision in the minimum measurement range.
For this purpose, relative humidity was set at 70% in the dynamic atmosphere system with an ammonia concentration of 14.4 mg/m^3. Higher humidity was not found to have an effect.

8.3 Limit of quantification

The limit of quantification was determined by replicate measurements. For this purpose, ammonia with a concentration of 0.09 mg/m^3 was vaporised in a dynamic test atmosphere system at a relative humidity of 40%. Nine sampling tubes were loaded with an air volume of 40 litres, processed and analysed. The mean variation at the defined limit of quantification was 7.3%.

8.4 Shelf-life

Shelf-life was determined at a low ($w = 1.0$ mg/m^3) and a high ($w = 10$ mg/m^3) ammonia concentration in the dynamic test atmosphere system. After sampling, the sample carriers were initially stored at room temperature for seven days and then kept in a re-

frigerator at +4 °C for a total storage period of up to four weeks. A total of nine duplicate determinations each were performed at regular intervals; no changes in ammonia
concentration were observed.

8.5 Interference

The present ion chromatography-based analytical method is specific for the ammonium
ion under the given operating conditions; alkali and alkaline earth cations are removed
and do not interfere with quantification. The column selected and the operating conditions described provide good resolution of the sodium and ammonium peaks.

8.6 Blank values

Blank values were checked by processing and analysing the sampling and backup sections from carbon bead tubes from different production batches.
Blank values were determined to be in the order of 0.2–0.3 mg/L for sample solution.
Because blank values may vary depending on supplier and production batch, they must
be determined individually for each batch that is used.

9 Discussion of the method

The blank values of the carbon bead tubes must be checked regularly.
When sampling ammonia, other ubiquitous particulate ammonium compounds are in-
variably also collected. Analytical differentiation is not possible.
The use of electrochemical background compensation instead of chemical suppression
yields similar results.
The present method has not been verified experimentally but a plausibility check was
performed on the basis of the available raw data.

10 References

[1] *Occupational Safety and Health Administration* (OSHA) (1991), Ammonia in Workplace At-
    mospheres – Solid Sorbent, Method No. ID-188, OSHA Analytical Laboratory
[2] *Deutsche Forschungsgemeinschaft* (2004) List of MAK und BAT Values 2004. Commission
    for the Investigation of Health Hazards of Chemical Compounds in the Work Area. Report
    No. 40. WILEY-VCH Verlag, Weinheim
[3] Technische Regel für Gefahrstoffe: Grenzwerte in der Luft am Arbeitsplatz – Luftgrenzwerte
    (TRGS 900), Ausgabe Oktober 2000, zuletzt geändert BArbBl. (2003) Nr. 4, S. 80

[4] *European Committee for Standardization* (CEN) (1994) EN 482 – Workplaceatmospheres – General requirements for the performance of procedures for the measurement of chemical agents. Brussels.

Authors: *D. Breuer, B. Heinrich*
Examiner: *R. Hebisch*

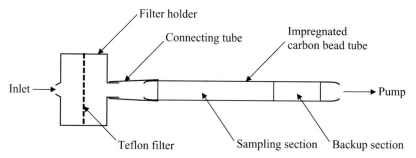

Fig. 1. Schematic representation of the combination sample carrier.

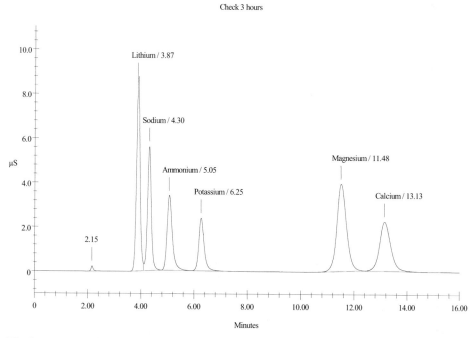

Fig. 2. Example of a chromatogram for cations, ammonia concentration 3.2 μg/mL.

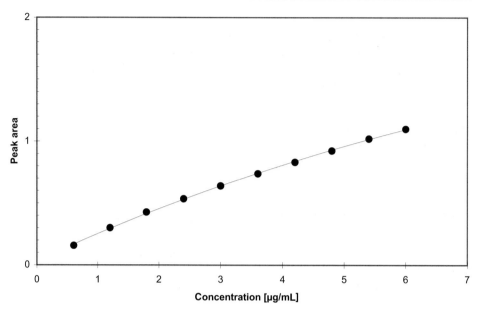

Fig. 3. Calibration curve for ammonia, concentration range 0.6–6.0 µg/mL.

**Federation of the Employment Accidents Insurance Institutions of Germany
(Hauptverband der Berufsgenossenschaften)
Centre for Accident Prevention and Occupational Medicine
Alte Heerstraße 111, 53757 Sankt Augustin**
Expert Committee Chemistry

Carcinogenic substances	Order number:	BGI 505-70E
Established methods	Issue:	April 2001

Method for the determination of 2-butanone oxime

Method tested and recommended by the Berufsgenossenschaften for the determination of 2-butanone oxime in working areas.
Both personal and stationary sampling can be conducted for the assessment of working areas.

Sampling with a pump and adsorption on Chromosorb 106
Gas chromatography after desorption
2-Butanone oxime-1-GC
(Issue: April 2001)

IUPAC name:	2-Butanone oxime
Synonyms:	Ethyl methyl ketoxime
	MEKO
CAS number:	96-29-7
Molecular formula:	C_4H_9NO
Molar mass (g/mol):	87.12

The MAK-Collection Part III: Air Monitoring Methods, Vol. 9. DFG, Deutsche Forschungsgemeinschaft
Copyright © 2005 WILEY-VCH Verlag GmbH & Co. KGaA, Weinheim
ISBN: 3-527-31138-6

Sampling with a pump and adsorption on Chromosorb 106

Summary

This method permits the determination of 2-butanone oxime concentrations in working areas averaged over the sampling time after personal or stationary sampling.

Principle:	A pump is used to draw a measured volume of air through a Chromosorb 106 tube. The adsorbed 2-butanone oxime is desorbed with methanol and determined by gas chromatography.

Technical data

Limit of
quantification: absolute: 0.8 ng 2-butanone oxime

relative: 0.2 mg/m^3 0.05 ml/m^3 (ppm) 2-butanone oxime for a 40-litre air sample, 10 ml desorption solution and 1 µl an injection volume.

Selectivity: In the presence of interfering components, the values determined may be too high. Interference can generally be eliminated by selecting a column with different separation characteristics.

Advantages: Personal sampling and selective determinations possible.

Disadvantages: No indication of peak concentrations.

Apparatus: Pump,
Gas meter or volumetric flow meter,
Chromosorb 106 tubes,
Gas chromatograph with nitrogen selective detector.

Detailed description of the method

Contents

1 Equipment, chemicals and solutions

1.1 Equipment

For sampling:
- Pump, suitable for flow rates of 0.33 L/min, e.g. GilAir-5, suppliers in Germany: DEHA Haan + Wittmer GmbH, D-71292 Friolzheim,
- Gas meter or volumetric flow meter,
- Adsorption tubes filled with Chromosorb 106 (standardised, consisting of an approx. 600 mg sampling section and an approx. 300 mg backup section of Chromosorb 106, separated by glass wool) e.g. from Supelco, Sigma-Aldrich Chemie GmbH, D-82024 Taufkirchen.

For sample preparation and analysis:
- Volumetric flasks, 10 mL
- Vials with screw caps and PTFE-coated septa, 20 ml nominal volume
- Pipette, 10 mL
- Syringes, 5 μL, 10 μL, 50 μL, 100 μL, 250 μL and 500 μL
- Disposable filter holder with PTFE-membrane 0.45 μm, e.g. Minisart SRP15, from Sartorius AG, D-37075 Göttingen
- Autosampler vials with screw caps

– Gas chromatograph with a nitrogen selective detector
– Data system

1.2 Chemicals and Solutions

Chemicals:
– 2-Butanone oxime 99%, e.g. from Lancaster Synthesis GmbH, D-65933 Frankfurt am Main, Catalogue No. 8196
– Methanol, analytical grade, e.g. from Merck, D-64271 Darmstadt, Catalogue No. 106009

Stock solution:
Solution of 0.369 mg/mL 2-butanone oxime in methanol
With a 10 µL syringe 4 µL of 2-butanone oxime is pipetted into a 10 ml volumetric flask containing approx. 8 mL methanol. The volumetric flask is then filled to the mark with methanol and shaken.

Calibration solutions:
Solutions of 0.74 µg/mL, 1.85 µg/mL, 3.69 µg/mL, 5.54 µg/mL or 7.38 µg/mL 2-butanone oxime in methanol
Syringes are used to inject 20, 50, 100, 150 and 200 µL of stock solution into five 10 ml volumetric flasks, each containing approx. 8 mL methanol. The flasks are then filled to the mark with methanol.
For a 40-litre air sample volume and 10 mL sample solution, these solutions cover a 2-butanone oxime concentration range of approx. 0.2 mg/m^3 to 2.0 mg/m^3 air.

Gases for gas chromatography:
– Helium, purity: 99.999%
– Hydrogen, purity: 99.999%
– Synthetic air, hydrocarbon-free
– Nitrogen, purity: 99.999%

2 Sampling

An adsorption tube is opened and connected to the pump such that the sampling section will be loaded first. The flow rate is set at 0.33 L/min. A sampling time of two hours then corresponds to an air sample volume of 40 litres. The pump and tube are carried by a person during working hours or used in a stationary position. On completion of sampling, the tube is tightly sealed.

3 Analytical determination

3.1 Sample preparation and analysis

The contents of the loaded Chromosorb tube are transferred to a 20 mL screw-cap vial and covered with 10 mL methanol and the vial is then capped. Desorption is complete after 16 hours. The extract is shaken and subsequently passed through a disposable filter holder into an autosampler vial (desorption solution).
Triplicate determinations are performed with 1 μL aliquots of desorption solution.
The quantitative evaluation of the chromatograms is performed by the external standard method.
To ensure that the desorption solution used and the Chromosorb 106 do not contain any interfering impurities, the filling of an unloaded Chromosorb 106 tube is desorbed with 10 mL methanol and 1 μL is injected into the gas chromatograph (blank value).

3.2 Operating conditions for gas chromatography

The method was characterised under the following experimental conditions:

Apparatus:	Siemens Sichromat 3 gas chromatograph with nitrogen selective detector and split/splitless injector
Column:	Fused silica capillary, stationary phase: Carbo Wax 20M-AM, length: 50 m, internal diameter: 0.25 mm, film thickness: 0.25 μm
Temperatures:	Furnace:
	Initial temperature: 70 °C, 1 minute isothermal,
	Heating rate: 10 °C/min until 180 °C,
	180 °C, 3 minutes isothermal,
	Injector: 200 °C
	Detector: 250 °C
Type of injection:	Split, 1:15
Carrier gas:	Helium, 1.5 mL/min
Detector gases:	Hydrogen, 1.7 mL/min,
	Synthetic air, 100 mL/min,
	Nitrogen (make-up gas), 30 ml/min
Injection volume:	1 μL

4 Calculations

4.1 Calibration

Aliquots of 1 µL of each of the calibration solutions described in Section 1.2 are injected into the gas chromatograph. The peak areas determined are plotted against the corresponding concentrations of 2-butanone oxime contained in the calibration solutions in order to construct the calibration curve. It is linear under the conditions described.

4.2 Calculation of the analytical result

The peak areas for 2-butanone oxime are determined, and the corresponding weight of analyte in the sample is read from the calibration curve in µg.
The 2-butanone oxime concentration by weight in the air sample is calculated in mg/m^3 according to equation (1):

$$c_w = \frac{w}{V \cdot \eta} \tag{1}$$

The concentration by volume c_v in mL/m^3 at 20 °C and 1013 hPa is calculated from c_w as follows (equation 2):

$$c_v = 0.277 \cdot c_w \tag{2}$$

Where:

c_w is the concentration by weight of 2-butanone oxime in the air sample, given in mg/m^3

c_v is the concentration by volume of 2-butanone oxime in the air sample, given in mL/m^3 (ppm)

w is the weight of 2-butanone oxime in the desorption solution in µg as determined from the appropriate calibration curve

V is the air sample volume in litres

η is the recovery.

5 Reliability of the method

5.1 Accuracy and recovery

The following spiking solutions were prepared in order to determine the relative standard deviation of the method:

Spiking solution: 40 µL 2-butanone oxime was dosed into a 10 mL volumetric flask containing approx. 8 mL methanol. The flask was then filled to the mark with methanol. The solution contained 36.9 mg 2-butanone oxime/10 mL methanol.

Aliquots of 2.2 µL, 5.4 µL and 325 µL were spiked onto individual adsorption tubes. Laboratory air (30–50% relative humidity) was subsequently drawn through each tube at a flow rate of 0.33 L/min for two hours. The solutions obtained after desorption were each injected three times into the gas chromatograph. This procedure covers the air concentrations given in the table below.

Six replicate determinations conducted according to the method described yielded the relative standard deviations and recoveries for 2-butanone oxime shown in the following table.

Concentration mg/m^3	Relative standard deviation %	Recovery
0.2	7.6	0.92
0.5	6.3	1.07
30	5.4	0.93

Based on this, the calculated mean recovery is 0.97.

5.2 Limit of quantification

The limit of quantification was estimated by loading six adsorption tubes per calibration point with different weights of 2-butanone oxime in the lower calibration range and then desorbed.

The desorption solutions were injected in triplicate into the gas chromatograph.

The mean values of the triplicate determinations were used to calculate the standard deviations for each set of six tubes.

A relative standard deviation of less than 10% was considered acceptable for the limit of quantification.

This was the case for an absolute 2-butanone oxime mass of 0.8 ng, or 8 µg 2-butanone oxime per tube.

Thus the relative limit of quantification is 0.2 mg/m^3 0.05 mL/m^3 (ppm) 2-butanone oxime for a 40-litre air sample, 10 mL desorption solution and an injection volume of 1 µL.

5.3 Selectivity

The selectivity of the method depends above all on the type of column used. The column specified here has proved reliable in practice. In the presence of interfering components, it may be necessary to use a column with different separation characteristics.

6 Discussion

The loaded tubes can be stored for 14 days at room temperature without loss of adsorbed 2-butanone oxime.

The characteristics of the method were determined with an older type of gas chromatograph. Accuracy and limit of quantification can be expected to improve through the use of more modern equipment, thus reducing the amount of experimental work needed (triplicate determinations).

7 References

2-Butanonoxim. In: BIA-Arbeitsmappe, Messung von Gefahrstoffen. Hrsg.: Berufsgenossenschaftliches Institut für Arbeitssicherheit-BIA, Sankt-Augustin, Blatt 6398, Verfahren Nr. 1. Erich Schmidt Verlag, Bielefeld, 22. Lieferung 1999

Author: *N. Lichtenstein*

2-Butenal

Method number   2

Application   Air analysis

Analytical principle Gas chromatography/mass spectrometry

Completed in   May 2003

Summary

The present method measures vaporous 2-butenal after adsorption on Chromosorb 106 and thermal desorption, using gas chromatography and mass spectrometry. Active sampling is performed. Quantitative evaluation is carried out with a test atmosphere of known concentration. There is a linear relationship between the peak areas and the concentrations of 2-butenal.

Characteristics of the method

Calculated from the calibration curve, according to DIN 32645

Accuracy:

Residual standard deviation: s_y = 6.98 ng ($\equiv$ 0.0349 mg/m^3)

Standard deviation of the method: s_{x0} = 2.27 ng ($\equiv$ 0.0114 mg/m^3)

Relative standard deviation of the method: V_{x0} = 1.68%

Calculation of the limit of quantification was performed with a chosen relative confidence interval of 25%.

Limit of quantification: 0.1 mg/m^3

Calculated from the tests performed in accordance with EN 482

Accuracy:

Relative standard deviation: s_{rel} = 6.4–9.8%
                      (at 0.09, 0.30, 0.94 and 2.3 mg/m^3)

The MAK-Collection Part III: Air Monitoring Methods, Vol. 9. DFG, Deutsche Forschungsgemeinschaft
Copyright © 2005 WILEY-VCH Verlag GmbH & Co. KGaA, Weinheim
ISBN: 3-527-31138-6

Mean variation:	$u = 16.1-24.9\%$
	(at 0.09, 0.30, 0.94 and 2.3 mg/m³)
Relative analytical uncertainty:	23.0–25.9%
Number of determinations:	$n = 6$
Recovery:	$\eta > 0.99$ (> 99%)
Limit of quantification:	0.038 mg/m³ (calculated from 10 blank values as the 12-fold standard deviation)
Sampling recommendation:	Sampling time: 0.5–2 hours
	Air sample volume: 0.2–0.5 litres

2-Butenal
[CAS Nos. 123-73-9 (*cis*-2-butenal), 4170-30-3 (*trans*-2-butenal)]

$CH_3 - CH = CH - CHO$ Synonyms: β-methylacrolein, crotonaldehyde

2-Butenal is a water-clear liquid with a strong odour. It exists in a *cis* and *trans* form, with the technical-grade product containing more than 95% *trans* isomer. Because 2-butenal has an aldehyde functional group and an olefinic double bond it is considered highly reactive. The compound is of technical importance as an intermediate in the manufacture of quinaldine, sorbic acid and butanol. The latter use, however, is increasingly declining in importance in favour of oxosynthesis.

The substance has been placed in Category 3 B in the List of MAK and BAT Values [1]. There is no longer a valid MAK value for 2-Butenal; instead, the Ausschuss für Gefahrstoffe (AGS; Hazardous Substances Committee) of the BMWA (Bundesministerium für Wirtschaft und Arbeit; German Federal Ministry of Economics and Labour) in 1995 established a technically based threshold limit value in air on the basis of the TRK concept and set it at 1 mg/m³ [2].

Author: *M. Tschickardt*
Reviewer: *W. Krämer*

2-Butenal

Method number	2
Application	Air analysis
Analytical principle	Gas chromatography/mass spectrometry
Completed in	May 2003

Contents

1 General principles

The present method measures vaporous 2-butenal after adsorption on Chromosorb 106 and thermal desorption, using gas chromatography and mass spectrometry. Active sampling is performed. Quantitative evaluation is carried out using a test atmosphere of known concentration. There is a linear relationship between the peak areas and the concentrations of the congeners.

2 Equipment, chemicals and solutions

2.1 Equipment

Adsorption tubes made of stainless steel, 6.3 mm × 90 mm, 5 mm internal diameter, packed with 300 mg Chromosorb 106 (e. g. from PerkinElmer Instruments)
Sampling pump, flow rate 5 mL/min (e. g. PP-1, supplied by Gilian, USA)
Gas chromatograph equipped with a thermal desorber (e. g. ATD-400, from PerkinElmer Instruments) and a flame ionisation detector
Mass spectrometer (e. g. Turbomass, from PerkinElmer Instruments)
Caps (e. g. Swagelok® with PTFE seals, PTFE or aluminium)
Capillary column DB-Wax 30 m, 0.5 μm film thickness; 0.25 mm internal diameter (e. g. from Promochem GmbH, D-46485 Wesel)
Gasmeter or stop clock and soap bubble flowmeter
If necessary, diffusion caps for passive sampling (e. g. PerkinElmer Instruments, Calatolgue No. 126433)
Dynamic test atmosphere generator according to VDI 3490, „Blatt 8" [3]

2.2 Chemicals

2-Butenal, for synthesis, purity >99%, from VWR, Catalogue No. 8.02667
Helium (carrier gas), purity 99.996%
Purified or synthetic air (free of hydrocarbons)
Hydrogen, purity 99.999%

2.3 Pretreatment of the adsorption tubes

Before use the adsorption tubes are heated at 170 °C in the thermal desorber and the blank values checked. For storage they are closed with Swagelok® or aluminium caps. Fitted with a diffusion cap, the tubes can also be used for passive sampling (see Figure 1) [4].

2.4 Calibration standards

With thermal desorption methods it is advisable to use test atmospheres for calibration. For this purpose, a 2-butenal test atmosphere is generated using a dynamic test atmosphere generation system (Figure 2).

2.5 Test atmospheres

Test atmospheres can be generated by various methods. One possibility is to generate test atmospheres by continuous injection [3]. 2-Butenal is continuously injected at 10 µL/h into a stream of complementary gas flowing at a rate of 100 mL/min in a dynamic test atmosphere generator at room temperature. At a second stage, 1 mL/min of this complementary gas is then diluted into a stream of dilution gas using a piston pump. Different concentrations are generated by setting the dilution gas flow at 500 or 1000 mL/min. Under the given laboratory conditions, the obtained test atmosphere concentrations were 1.41 and 2.83 mg/m^3. The apparatus set-up is shown in Figure 2. The adsorption tubes are loaded with known volumes of the generated test atmosphere and used for calibration.

3 Sampling and sample preparation

Sample collection can be performed as stationary or personal sampling. The adsorption tubes must be heated in the thermal desorber before sampling because after prolonged periods of storage interfering substances released from the sealing material of the PTFE caps can be adsorbed on the collection material. The adsorption tubes are opened at the beginning of sampling. The parameters which are important for the determination of the concentrations in air (sample volume, temperature, atmospheric pressure and relative humidity) are recorded in a sampling record.
Sampling is carried out in the breathing zone for personal sampling. The opening of the adsorption tube should not be obstructed.
After sampling, the adsorption tubes are closed with PTFE caps. These samples should be analysed immediately. If the samples are to be stored for a longer period of time until analysis, the tubes must be sealed with Swagelok$^®$ screw caps with PTFE seals.
The method has been tested and found usable at relative humidities from 5 to 80%.

3.1 Active sampling

A sampling pump is used to draw the air to be analysed through the adsorption tube continuously at a flow rate of 5 mL/min over a period of 30 minutes to two hours. On completion of sampling, the loaded adsorption tube is sealed with caps at both ends.

3.2 Passive sampling

Before sampling, the cap at the end of the tube intended for passive sampling is removed and replaced by a diffusion cap (Figure 1). A sampling time of four to eight hours is recommended. After the end of sampling, the loaded adsorption tube is sealed with caps at both ends.

For the use of diffusive samplers and their limitations see [4] and [5].

4 Operating conditions

Apparatus:	Autosystem XL gas chromatograph
Column:	Material: Quartz capillary
	Length: 30 m
	Internal diameter: 0.25 mm
	Stationary phase: DB-WAX
	Film thickness: 0.5 μm
Flow splitting of the eluate:	Split ratio via splitter capillaries at the column outlet: set at 5:1 (MS/FID)
Detectors:	Quadrupole mass spectrometer (e.g. Turbomass, from PerkinElmer Instruments) and flame ionisation detector (FID) connected to the splitter capillaries
Detector temperature:	320 °C
Detector gases:	Hydrogen (purity 99.999%), synthetic air
Detector gas flow rates:	45 or 450 mL/min
Temperature programme:	50 °C (10 min) $\xrightarrow{8\,°C/min}$ 120 °C (1.2 min) $\xrightarrow{12\,°C/min}$ 200 °C (10 min)

Thermal desorption

Apparatus:	ATD-400 (from PerkinElmer Instruments)
Desorption temperature:	170 °C
Desorption time:	10 minutes
Transfer line:	200 °C
Length of transfer line:	1.5 m
Cold trap (adsorption):	20 °C
Cold trap (injection):	325 °C
Cold trap packing:	38 mg Carbotrap and 23 mg Carbosieve S-III (e.g. PerkinElmer Instruments, Air Monitoring Trap, Catalogue No. L4275108)
Carrier gas:	Helium, purity 99.996%
Carrier gas pressure:	125 kPa
Input split:	closed, 0 mL/min
Desorb flow:	30 mL/min
Output split:	5 mL/min

Figure 3 shows a chromatogram obtained under the conditions given above.

Operating conditions for mass spectrometry

Apparatus:	Turbomass, from PerkinElmer Instruments	
Temperatures:	Ion source:	180 °C
	Transfer line:	180 °C
Type of ionisation:	Electron impact	
Pressure inside the ion source:	3×10^{-6} kPa	
Electron current:	50 µA	
Ionisation energy:	70 eV	
Mass range:	35–300 amu	
Quantification mass:	35–300 amu	
Rate:	1 scan/s	
Interscan delay:	0.1 s	

5 Analytical determination

The adsorption tubes are heated in a compatible thermal desorber, and the adsorbed components are transferred to a packed cold trap with a carrier gas. After complete desorption the split outlet is opened and the cold trap heated. The sample reaches the column as a narrow band. The thermal desorber is connected to gas chromatograph via a deactivated quartz capillary. The instrument settings have to be modified if other types of thermal desorbers are used. After setting up the thermal desorber and the gas chromatograph (see Section 4) the calibration standards and the samples are analysed.

6 Calibration

The amount of the calibration standards loaded should be equivalent to concentrations between 0.1 and twice the threshold limit value in air [6]. Aliquots of 5, 25, 50, 100, 200 and 250 mL of the test atmosphere generated as described in Section 2.5 are drawn through adsorption tubes.

Table 1. Calibration weights (ng 2-butenal per adsorption tube).

Aliquots (test atmosphere concentration 1.41 mg/m³) mL	Aliquots (test atmosphere concentration 2.83 mg/m³) mL	Loading ng	Concentration for a sample volume of 200 mL mg/m³
5		7.06	0.0353
25		35.32	0.1766
50		70.64	0.3532
100		141.28	0.7064
200		282.57	1.4128
250		353.21	1.7660
	250	706.42	3.5321

The calibration curve is constructed by plotting the peak areas for atomic mass $m/z = 69$ as determined by the MS data system against the corresponding load in ng (Figure 3). The calibration function is linear in the specified range and should be checked regularly in routine analysis. For this purpose, a sample of test atmosphere of known concentration should be analysed with each analytical series.

7 Calculation of the analytical result

Using the peak areas obtained, the corresponding weight X in ng is read from the calibration curve. The corresponding concentration by weight (ρ) is calculated according to the following equation:

$$\rho = \frac{X}{V} \tag{1}$$

At 20 °C and 1013 hPa:

$$\rho_0 = \rho \times \frac{273 + t_a}{293} \times \frac{1013}{p_a} \tag{2}$$

Where:
ρ is the concentration of a component in mg/m³
ρ_0 is the concentration in mg/mg³ at 20 °C and 1013 hPa
X is the weight of the component in the sample in ng
t_a is the temperature during sampling in °C
p_a is the atmospheric pressure during sampling in hPa
V is the air sample volume in mL (calculated from the flow rate and the sampling time)

For passive sampling:

$$V = U_m \times t \tag{3}$$

$$U_m = \frac{60 \times D_1 \times A}{Z} \tag{4}$$

Where:
U_m is the sampling rate in mL/min
t is the sampling time (min)
D_1 is the diffusion coefficient in air (cm^2/s), calculated here as 0.0882 cm^2/s at 25 °C
        and 1013 hPa [7, 8]
A is the surface of the sampler (cm^2) (ATD type: 0.196 cm^2)
Z Diffusion distance (cm) (ATD type: 1.5 cm)

The rest of the calculation is carried out according to (1). The concentration by volume σ in mL/m^3 is calculated as follows:

$$\sigma = 0{,}343 \times \rho_0 \tag{5}$$

8 Reliability of the method

The characteristics of the method were determined according to the standards EN 482 and DIN 32645 [9]. For this purpose, test atmospheres were generated which contained concentration levels of 0.085 mg/m^3, 0.298 mg/m^3, 0.939 mg/m^3 and 2.31 mg/m^3. The test atmospheres were adjusted with water to 70–89% relative humidity. For each of these concentrations, 6 samples (sample volumes of 200 or 500 mL) were drawn at room temperature. The adsorption tubes were sealed with Swagelok$^{®}$ caps, stored at room temperature and analysed after 4 weeks. Further statistical data according to DIN 32645 were determined using the pairs of calibration values.

8.1 Accuracy

8.1.1 Precision

In order to determine precision, three humidified test atmospheres containing different concentrations were generated. Per test atmosphere, six calibration samples of 200 mL volume were drawn and stored at room temperature for four weeks. The samples were subsequently analysed as described in Section 5. The following data were obtained:

Table 2. Relative standard deviation s_{rel}, $n = 6$ determinations.

Test atmosphere concentration	Loading	Mean value	Relative standard deviation s_{rel}	Mean variation u
mg/m^3	ng	ng	%	%
0.298	59.5	62.8	8.9	23.0
0.939	187.7	159.6	6.4	25.9
2.31	461.2	433.5	9.8	24.5

Accuracy from the calibration curve according to DIN 32645

Residual standard deviation: s_y = 6.98 ng ($\equiv$ 0.0349 mg/m^3)
Standard deviation of the method: s_{x0} = 2.27 ng ($\equiv$ 0.0114 mg/m^3)
Relative standard deviation of the method: V_{x0} = 1.68%

The recovery was tested by twice heating various tubes. The concentrations determined were in the range between 0.3 and 2.4 mg/m^3. The recovery for the specified concentrations and a single heating is over 0.99 (> 99%).

With thermal desorption methods and simultaneous calibration by means of test atmospheres, it is not possible to determine recovery.

Data from the shelf-life tests according to EN 482

Relative standard deviation: s_{rel} = 6.4–9.8%
                  (0.09, 0.30, 0.94 and 2.3 mg/m^3)
Mean variation: u = 16.1–24.9%
                  (0.09, 0.30, 0.94 and 2.3 mg/m^3)
Relative analytical uncertainty: 23.0–25.9%
Number of determinations: n = 6

8.1.2 Accuracy of the mean

In order to determine the accuracy of the mean, humidified test atmospheres containing different concentrations were generated. Per test atmosphere, six calibration samples of 200 mL or 500 mL volume were drawn and stored at room temperature for four weeks. The samples were subsequently analysed as described in Section 5. The following data were obtained:

Table 3. Recovery, $n = 6$ determinations.

Test atmosphere concentration	Sample volume	Loading	Mean value	Standard deviation s_{rel}	Recovery
mg/m^3	mL	ng	ng	ng	%
0.085	500	42.9	41.6	2.8	97.0
0.298	200	59.5	62.8	5.6	105.5
0.939	200	187.7	159.6	10.3	85.0
2.31	200	461.2	433.5	42.7	94.0

Mean recovery $\eta = 0.95$ (95%)

8.2 Limit of quantification

Calculation of the limit of quantification was performed according to DIN 32645 with a chosen relative confidence interval of 25%. A level of 0.1 mg/m^3 was calculated from the calibration curve with an assumed air sample volume of 200 mL.
The limit of quantitation was checked by calculation from the analytical results for 10 blanks and ascertained as 0.038 mg/m^3 (12-fold standard deviation, air sample volume 200 mL).

8.3 Shelf-life

Shelf-life tests of the loaded sample carriers were performed over a period of four weeks. Storage was at room temperature. No significant losses were detected (Table 3).

8.4 Interference

Determination of 2-butenal using the specified column is susceptible to interference from toluene. Therefore, a mass spectrometer is used as the detector. If the presence of relevant levels of toluene can be excluded it is sufficient to use a flame ionisation detector.
2-Butenal occurs as a mixture of its *cis* and *trans* isomers. The integration settings for the method of calculating the results need to be adjusted accordingly. In the present case, addition of the two peaks was used for method validation.

8.5 Capacity of the adsorbent

The capacity of the adsorption tube is sufficient to allow determination of 2-butenal concentration levels of up to 2.3 mg/m^3.

9 Discussion of the method

The present analytical method permits the determination of 2-butenal in workplace air. Testing was additionally performed at a concentration level of 0.085 mg/m^3 using an air sample volume of 500 mL. The characteristics of the method were confirmed.

When the specified separation capillary is used together with a flame ionisation detector there is interference from toluene.

Assessment of the passive sampling method was accomplished by generating a test atmosphere containing 1.62 mg/m^3 and exposing six passive samplers to the test atmosphere for four hours. Quantitative evaluation was carried out using a calculated diffusion coefficient of 0.0882 cm^2/sec, corresponding to a sampling rate of 0.7056 mL/min. Based on the specified sampling rate, quantitative evaluation yielded a recovery of 80.2% with a relative standard deviation of 3.2%.

10 References

[1] *Deutsche Forschungsgemeinschaft* (2004) List of MAK und BAT Values 2004. Commission for the Investigation of Health Hazards of Chemical Compounds in the Work Area. Report No. 40. Wiley-VCH Verlag, Weinheim.

[2] Technische Regel für Gefahrstoffe: Grenzwerte in der Luft am Arbeitsplatz – Luftgrenzwerte (TRGS 900), Ausgabe Oktober 2000, zuletzt geändert BArbBl. (2003) No. 4, p. 80.

[3] *Verein Deutscher Ingenieure*: VDI-Richtlinie 3490, Blatt 1–16.

[4] *Blome H.* (1988) Möglichkeiten und Grenzen der Verwendung von Diffusionssammlern zur Probenahme gas- und dampfförmiger Stoffe in der Luft in Arbeitsbereichen, Staub Reinhalt. Luft 48: 177–181.

[5] *Health and Safety Executive, Committee of analytical requirements, Working Group 5* (2001) Diffusive uptake rates on the Perkin Elmer diffusive tube, The Diffusive Monitor 12: 6–9.

[6] *Europäisches Komitee für Normung* (CEN) (1994) EN 482 – Arbeitsplatzatmosphäre – Allgemeine Anforderungen an Verfahren zur Messung von chemischen Arbeitsstoffen. Brüssel. Beuth Verlag, Berlin.

[7] *Ullmans Encyklopädie der technischen Chemie* (1972) Ermittlung von Diffusionskoeffizienten, Band 1: 149–150.

[8] *Nelson, Gary O.* (1992) Gas mixtures: preparation and control. Lewis Publishers, ISBN 0-87371-298-6.

[9] *Deutsches Institut für Normung e.V.* (DIN) (1994) DIN 32645 – Chemische Analytik – Nachweis-, Erfassungs- und Bestimmungsgrenze. Beuth Verlag, Berlin.

Author: *M. Tschickardt*
Reviewer: *W. Krämer*

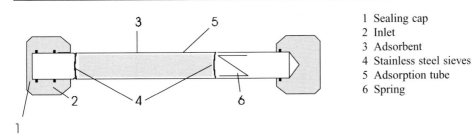

Fig. 1. Adsorption tube made of stainless steel, 6.3 mm × 90 mm, 5 mm internal diameter.

1 Sealing cap
2 Inlet
3 Adsorbent
4 Stainless steel sieves
5 Adsorption tube
6 Spring

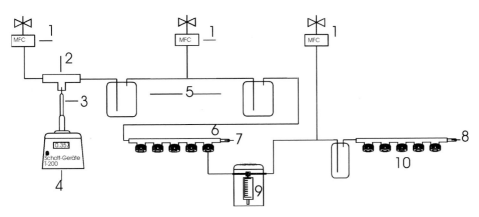

1 Pressure control for zero gas
2 Injector
3 Piston
4 Piston burette
5 Buffer vessel

6 Sampling manifold for primary gas
7 Excess primary gas
8 Excess calibration test atmosphere
9 Piston pump
10 Sampling manifold for calibration test atmosphere

Fig. 2. Dynamic test atmosphere generation system.

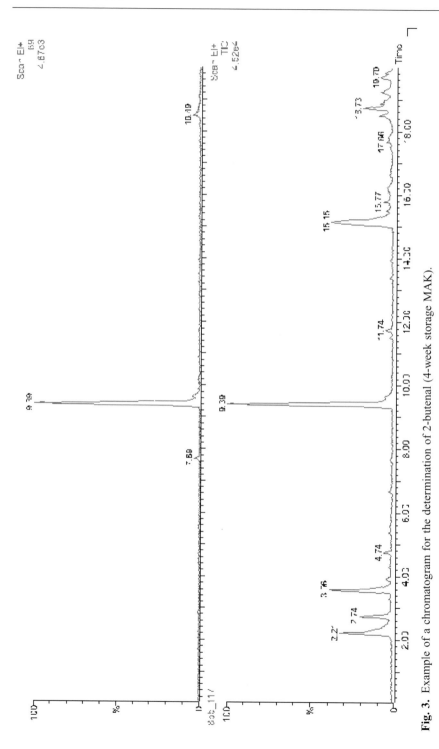

Fig. 3. Example of a chromatogram for the determination of 2-butenal (4-week storage MAK).

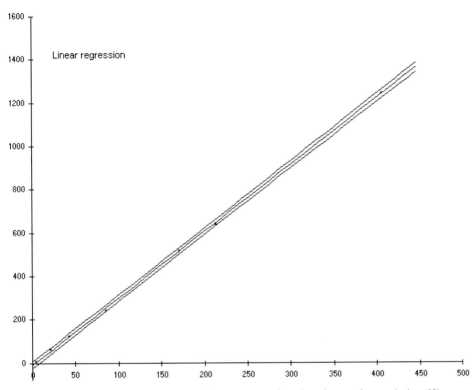

Fig. 4. Calibration curve for 2-butenal with nanograms plotted against peak area (m/z = 69).

Dicyclopentadiene

Method number	1
Application	Air analysis
Analytical principle	Gas chromatography/flame ionisation detector
Completed in	November 2003

Summary

The present method measures vaporous dicyclopentadiene after adsorption on Chromosorb 106 and thermal desorption, using a gas chromatograph equipped with a flame ionisation detector. Active sampling is performed. Quantitative evaluation is carried out using a test atmosphere of known concentration. There is a linear relationship between the peak areas and the concentrations of dicyclopentadiene in the concentration range investigated.

Characteristics of the method

Accuracy:

Standard deviation (rel.)	$s_{rel} =$	4.4%
Mean variation	$u\ \ =$	15.5%
	for $n = 6$ determinations and $c = 0.57$ mg/m^3	
Standard deviation (rel.)	$s_{rel} =$	5.1%
Mean variation	$u\ \ =$	11.6%
	for $n = 6$ determinations and $c = 2.0$ mg/m^3	
Standard deviation (rel.)	$s\ \ =$	3.6%
Mean variation	$u\ \ =$	10.0%
	for $n = 6$ determinations and $c = 4.4$ mg/m^3	

Recovery: $\quad \eta > 0.99\ (> 99\%)$

Limit of quantification: $\quad 0.082$ mg/m^3

Sampling recommendation: Sampling time: 0.5 hours
Air sample volume: 0.2 litres

The MAK-Collection Part III: Air Monitoring Methods, Vol. 9. DFG, Deutsche Forschungsgemeinschaft
Copyright © 2005 WILEY-VCH Verlag GmbH & Co. KGaA, Weinheim
ISBN: 3-527-31138-6

Dicyclopentadiene [CAS No. 77-73-6]

$C_{10}H_{12}$

Synonyms: Cyclopentadiene dimer; 3a,4,7,7a-tetrahydro-4,7-methano-1H-indene

Dicyclopentadiene is a water-clear liquid with an odour of camphor even at low concentration levels. It is readily soluble in organic solvents and serves practically as a storage form of the highly reactive compound cyclopentadiene, which is easily obtained by distillation (bp 170 °C; depolymerisation).

Dicyclopentadiene is used inter alia in the manufacture of impact-resistant plastics, which requires no fibre reinforcement. Processing requires no releasing or separating agents and the products obtained can be easily coated with paints, lacquers or varnishes.

The Deutsche Forschungsgemeinschaft proposes for cyclopentadiene a threshold limit value of 2.7 mg/m^3 (0.5 mL/m^3) in air [1] and placement in peak limitation category I(1). Dicyclopentadiene has been classed in Group IIc in the 2004 List of MAK and BAT Values [1] for assessment of pregnancy risks.

The present method was checked for plausibility by experts from the DFG working group on air analyses.

Author: *M. Tschickardt*
Examiners: *M. Ball, M. R. Lahaniatis*

Dicyclopentadiene

Method number   1

Application   Air analysis

Analytical principle   Gas chromatography/flame ionisation detector

Completed in   November 2003

Contents

1 General principles

The present method measures vaporous dicyclopentadiene after adsorption on Chromosorb 106 and thermal desorption, using a gas chromatograph equipped with a flame ionisation detector. Active sampling is performed. Quantitative evaluation is carried out using a test atmosphere of known concentration. There is a linear relationship between the peak areas and the concentrations of dicyclopentadiene in the concentration range investigated.

2 Equipment, chemicals and solutions

2.1 Equipment

Adsorption tubes made of stainless steel, 6.3 mm × 90 mm, 5 mm internal diameter, prepacked with 300 mg Chromosorb 106 (e.g. from Supelco or PerkinElmer Instruments)
Sampling pump, flow rate 5 mL/min (e.g. PP-1, supplied by Gilian, USA)
Gas chromatograph equipped with a thermal desorber (e.g. ATD 400, from PerkinElmer Instruments) and a flame ionisation detector
Caps (e.g. Swagelock® with PTFE seals, PTFE or aluminium)
DB-5 capillary column, 30 m, 0.5 μm film thickness, 0.25 mm internal diameter (e.g. from Promochem GmbH, Wesel)
Gasmeter or stop clock and soap bubble flowmeter and, if necessary, diffusion caps for passive sampling (e.g. PerkinElmer Instruments, Calatolgue No. 126433)

2.2 Chemicals

Dicyclopentadiene, stabilised, for synthesis, purity > 99%, from Merck, Catalogue No. 8.03038
Purified or synthetic air (free of hydrocarbons)
Hydrogen, purity 99.999%
Helium (carrier gas), purity 99.996%
Chromosorb 106, 60–80 mesh

2.3 Pretreatment of the adsorption tubes

The adsorption tubes come prepacked with 300 mg Chromosorb 106 and do not require any special pretreatment. Fitted with a diffusion cap, the adsorption tubes can also be used for passive sampling (see Figure 1) [2]. Before use the adsorption tubes are heated at 170 °C in the thermal desorber and the blank values checked. For storage they are closed with Swagelok® or aluminium caps.

2.4 Calibration standards

With thermal desorption methods it is advisable to use test atmospheres for calibration. Undiluted dicyclopentadiene is used to generate a test atmosphere using a two-stage test atmosphere generator.

2.5 Test atmospheres

Test atmospheres can be generated by various methods [3]. One possibility is to generate test atmospheres by continuous injection [4].

Dicyclopentadiene test atmosphere (calibration test atmosphere) is generated in a dynamic test atmosphere generator by continuously injecting pure dicyclopentadiene at 40 µL/h into a stream of air flowing at a rate of 230 mL/min (primary gas). This yields a concentration of 2831 mg/m^3. This primary gas is continuously diluted in air by means of a piston pump (yielding the calibration test atmosphere). A dosing rate of 1 mL primary gas per minute and zero gas flow rates of 0.65, 1.40 or 4.98 L/min yield respective concentrations of 4.4, 2.04 and 0.57 mg dicyclopentadiene per m^3 of ready-to-use calibration test atmosphere. The apparatus set-up is shown in Figure 2. The adsorption tubes are loaded with known volumes of the generated test atmosphere/calibration test atmosphere and used for calibration.

3 Sampling and sample preparation

Sample collection can be performed as stationary or personal sampling. The adsorption tubes must be heated in the thermal desorber before sampling because after prolonged periods of storage interfering substances released from the sealing material of the PTFE caps can be adsorbed on the collection material. At the beginning of sampling, the adsorption tube is opened and connected to a pump or used as a diffusive sampler. The parameters which are important for the determination of the concentrations in air (sample volume or sampling time, temperature, atmospheric pressure and relative humidity) are recorded in a sampling record.

Sampling is carried out in the breathing zone. The opening of the adsorption tube should not be obstructed.

After sampling, the adsorption tubes are capped with Swagelok® screw caps with PTFE seals.

3.1 Active sampling

The sampling pump is used to draw the air to be analysed through the adsorption tube at a flow rate of 5 mL/min. On completion of sampling, the loaded adsorption tube is sealed with caps at both ends.

3.2 Passive sampling

Before sampling, the cap at the end of the tube intended for passive sampling is removed and replaced by a diffusion cap (Figure 1). A sampling time of 4 to 8 hours is recommended. On completion of sampling, the loaded adsorption tube is sealed with caps at both ends. For the use of diffusive samplers and their limitations see [2] and [5].

4 Operating conditions

Apparatus:	Autosystem XL gas chromatograph (from PerkinElmer Instruments)	
Column:	Material:	Quartz capillary
	Length:	30 m
	Internal diameter:	0.25 mm
	Stationary phase:	DB-5
	Film thickness:	0.5 μm
Detectors:	Flame ionisation detector (FID)	
Detector temperature:	250 °C	
Detector gases:	Hydrogen (purity 99.999 %), synthetic air	
Detector gas flow rates:	45 or 450 mL/min	
Temperature programme:	$50\,°C$ (10 min) $\xrightarrow{8\,°C/min}$ $120\,°C$ (1.2 min)	

Thermal desorption:

Apparatus:	ATD-400 (from PerkinElmer Instruments)
Desorption temperature:	170 °C
Desorption time:	10 minutes
Transfer line:	200 °C
Length of transfer line:	1.5 m
Cold trap (adsorption):	20 °C
Cold trap (injection):	325 °C
Cold trap packing:	38 mg Carbotrap and 23 mg Carbosieve S-III (e. g. PerkinElmer Instruments, Air Monitoring Trap, Catalogue No. L4275108)
Carrier gas:	Helium, purity 99.996 %
Carrier gas pressure:	125 kPa
Input split:	closed, 0 mL/min
Desorb flow:	30 mL/min
Output split:	5 mL/min

5 Analytical determination

The adsorption tubes are heated in a compatible thermal desorber, and the adsorbed components are transferred to a packed cold trap with a carrier gas. After complete desorption the split outlet is opened and the cold trap heated. The sample reaches the column as a narrow band. The thermal desorber (ATD 400) is connected to gas chromatograph via a deactivated quartz capillary.

The instrument settings have to be modified if other types of thermal desorbers are used. After setting up the thermal desorber and the gas chromatograph (see Section 4) the calibration standards and the samples are analysed (Figure 3).

6 Calibration

The amount of the calibration standards loaded should be equivalent to concentrations between 0.1 and twice the threshold limit value. Aliquots of 1, 5, 25, 50, 100, 200, 300, 400 and 500 mL of the test atmosphere generated as described in Section 2.5 are drawn through adsorption tubes (Table 1).

Table 1. Calibration weights (ng dicyclopentadiene per adsorption tube).

Aliquots (test atmosphere concentration 1.97 mg/m^3) mL	Loading ng	Concentration for a sample volume of 200 mL mg/m^3
1	1.97	0.010
5	9.86	0.049
10	19.7	0.099
25	49.3	0.247
50	98.9	0.495
100	197	0.986
200	394	1.97
300	591	2.96
400	789	3.95
500	986	4.93

The calibration curve is constructed by plotting the peak areas determined by the data system against the corresponding loading in ng (Figure 4). The calibration curve is linear in the range indicated here. The calibration curve should be checked regularly in routine analysis. For this purpose, a sample of test atmosphere of known concentration should be analysed with each analytical series.

7 Calculation of the analytical result

Using the peak areas obtained, the corresponding weight X in ng is read from the calibration curve. The corresponding concentration by weight (ρ) is calculated according to the following equation:

$$\rho = \frac{X}{V} \tag{1}$$

At 20 °C and 1013 hPa:

$$\rho_0 = \rho \times \frac{273 + t_a}{293} \times \frac{1013}{p_a} \tag{2}$$

where:
ρ is the concentration of the component in mg/m³
ρ_0 is the concentration in mg/mg³ at 20 °C and 1013 hPa
X is the weight in ng of the component in the analytical sample
t_a is the temperature during sampling in °C
p_a is the atmospheric pressure during sampling in hPa
V is the air sample volume in ml (calculated from the flow rate and the sampling time)

For passive sampling:

$$V = U_m \times t \tag{3}$$

$$U_m = \frac{60 \times D_1 \times A}{Z} \tag{4}$$

where:
U_m is the sampling rate in mL/min, in this case 0.46 mL/min
t is the sampling time (min)
D_1 is the diffusion coefficient in air (cm²/s), which here is 0.0575 cm²/s according to [6]
A is the surface of the sampler (cm²) (ATD type: 0.196 cm²)
Z Diffusion distance (cm) (ATD type: 1.5 cm)

The rest of the calculation is carried out according to Equation (1).
The concentration by volume σ in mL/m³ is calculated as follows:

$$\sigma = 0.182 \times \rho_0 \tag{5}$$

8 Reliability of the method

The characteristics of the method were determined according to the standard EN 482 [7]. For this purpose, 3 test atmospheres were generated which contained concentration levels of 0.57 mg/m^3, 2.0 mg/m^3 and 4.4 mg/m^3. The test atmospheres were adjusted with water to 40 % relative humidity. For each of these concentrations, 6 samples (sample volume 200 mL) were drawn at 24 °C. The adsorption tubes were sealed with Swagelok® caps and analysed. Additionally, 12 samples were drawn at a concentration level of 2.1 mg/m^3, 6 of which were analysed after one day and the other 6 after 22 days of storage at room temperature.

A comparison between active and passive sampling was performed by personal and stationary sampling in the context of workplace measurements.

8.1 Accuracy

8.1.1 Precision

In order to determine precision, three humidified test atmospheres containing different concentrations were generated. Per test atmosphere, six calibration samples of 200 mL volume were drawn. The samples were subsequently analysed as described in Section 5. The following data were obtained (Table 2):

Table 2. Standard deviation s_{rel}, $n = 6$ determinations.

Test atmosphere concentration	Mean value	Relative standard deviation	Relative analytical uncertainty
mg/m^3	ng	s_{rel} %	u %
0.57	0.612	4.4	15.5
2.0	2.067	5.1	11.6
4.4	4.493	3.6	10.0

Recovery: The recovery was tested by twice heating various tubes. The initial concentrations determined were in the range between 0.57 and 4.4 mg/m^3. The recovery for the specified concentrations was over 0.99 (> 99 %).

8.1.2 Accuracy of the mean

With thermal desorption methods and simultaneous calibration by means of test atmospheres, it is not possible to determine recovery.

Active and passive samples were taken in parallel over a period of two hours in the context of workplace measurements. A room temperature of 20 °C and a relative humidity

of 79% were measured during sampling. Calculation of the analytical results yielded
the following pairs of values (Table 3):

Table 3. Results for comparative measurements with active and passive sampling.

Sample number	Active sampling: concentration mg/m^3	Passive sampling: concentration mg/m^3
1	6.6	6.4
2	72	86
3	6.3	6.7
4	2.3	2.4
5	2.1	2.3

The mean recovery for passive sampling is 107% in the concentration range investigated.

8.2 Limit of quantification

The limit of quantification was determined from the analytical results for 10 samples
with a concentration of 0.082 mg/m^3. A value of 0.094 mg/m^3 was determined with a
standard deviation of 0.005 mg/m^3. Hence, the stated concentration was defined as the
limit of quantification.

8.3 Shelf-life

A humidified test atmosphere of known concentration was generated. Per test atmosphere, 12 calibration samples of 200 mL volume were drawn and six samples were
stored at room temperature for 22 days. The samples were subsequently analysed as described in Section 5. The data shown in Table 4 were obtained.

Table 4. Shelf-life tests with analysis after 1 or 22 days, $n = 6$ determinations.

Test atmosphere concentration mg/m^3	Mean value after one day mg/m^3	Standard deviation s ng	Mean value after 22 days mg/m^3	Standard deviation s ng	Recovery %
2.1	2.09	0.03	1.92	0.041	92.0

8.4 Interference

The determination of dicyclopentadiene is carried out on an apolar column. There is interference from limonene as a coeluting compound. Prepacked adsorption tubes gave no blank value.

8.5 Capacity of the adsorbent

The capacity of the adsorption tube is sufficient to allow determination of more than 880 ng dicyclopentadiene, corresponding to a concentration of 4.4 mg/m^3.

9 Discussion of the method

The present analytical method permits the determination of dicyclopentadiene in workplace air by active or passive sampling. The specified column also permits parallel determination of other components in the air. Interferences cannot be excluded in such cases. Limonene is a known interfering component. In this case, the additional use of a mass spectrometer provides sufficient selectivity. The validation parameters were obtained using test atmospheres with a relative humidity of 40%, while comparison measurements at a dicyclopentadiene-processing plant were taken at 80% humidity.

10 References

[1] *Deutsche Forschungsgemeinschaft* (2004) List of MAK und BAT Values 2004. Commission for the Investigation of Health Hazards of Chemical Compounds in the Work Area. Report No. 40. WILEY-VCH Verlag, Weinheim

[2] *H. Blome* (1988) Möglichkeiten und Grenzen der Verwendung von Diffusionssammlern zur Probenahme gas- und dampfförmiger Stoffe in der Luft in Arbeitsbereichen, Staub Reinhaltung der Luft 48, 177–181

[3] *Greim H.* (1994), Analytische Methoden zur Prüfung gesundheitsschädlicher Arbeitsstoffe – Luftanalysen, Spezielle Vorbemerkungen 1, 38–67

[4] *Verein Deutscher Ingenieure* (Hrsg.) (1981) VDI-Richtlinie 3490 Blatt 8, Prüfgase-Herstellung durch kontinuierliche Injektion.

[5] *Health and Safety Executive* (eds.) (2002) Diffusive uptake rates on the Perkin Elmer diffusive tube. The Diffusive Monitor 13

[6] *K.-H. Pannwitz* (1983) Diffusionskoeffizienten, Drägerheft 327, 6–13 Drägerwerk, Lübeck

[7] *Europäisches Komitee für Normung* (CEN) (1994) EN 482 – Arbeitsplatzatmosphäre – Allgemeine Anforderungen an Verfahren zur Messung von chemischen Arbeitsstoffen. Brüssel. Beuth Verlag, Berlin.

Author: *M. Tschickardt*
Examiners: *M. Ball, M. R. Lahaniatis*

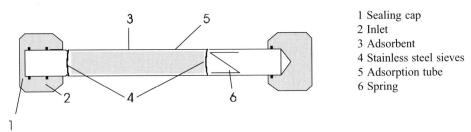

Fig. 1. Adsorption tube made of stainless steel, 6.3 mm × 90 mm, 5 mm internal diameter.

1 Sealing cap
2 Inlet
3 Adsorbent
4 Stainless steel sieves
5 Adsorption tube
6 Spring

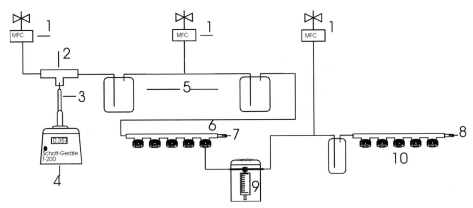

1 Pressure control for zero gas
2 Injector (80 °C)
3 Piston
4 Piston burette
5 Buffer vessel

6 Sampling manifold for primary gas
7 Excess primary gas
8 Excess calibration test atmosphere
9 Piston pump
10 Sampling manifold for calibration test atmosphere

Fig. 2. Dynamic test atmosphere generation system.

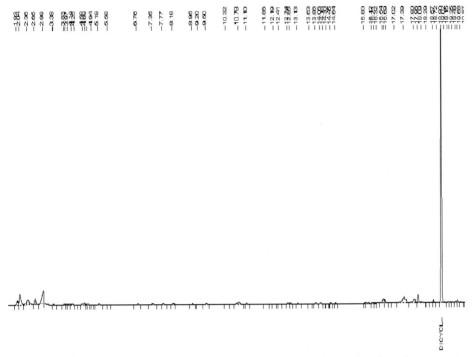

Fig. 3. Example of a chromatogram for the determination of dicyclopentadiene (concentration 2.1 mg/m^3).

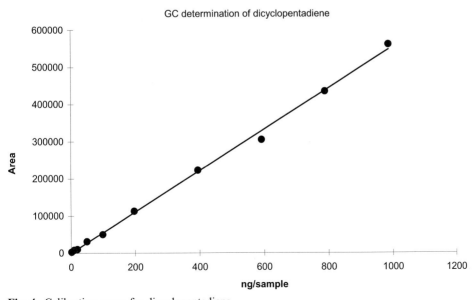

Fig. 4. Calibration curve for dicyclopentadiene.

Federation of the Employment Accidents Insurance Institutions of Germany
(Hauptverband der Berufsgenossenschaften)
Centre for Accident Prevention and Occupational Medicine
Alte Heerstraße 111, 53757 Sankt Augustin
Expert Committee Chemistry

Carcinogenic substances	Order number:	BGI 505-67E
Established methods	Issue:	September 1999

Method for the determination of 2,4-dinitrotoluene, 2,6-dinitrotoluene and 2,4,6-trinitrotoluene

Method tested and recommended by the Berufsgenossenschaften for the determination of 2,4-dinitrotoluene, 2,6-dinitrotoluene and 2,4,6-trinitrotoluene in working areas after discontinuous sampling.
Both personal and stationary sampling can be conducted for the assessment of working areas.

Sampling with a pump and adsorption on a combined sample holder consisting of a quartz fibre filter and a Tenax adsorption tube; high-performance liquid chromatography (HPLC) after elution.
„Di- and trinitrotoluenes-2-HPLC"
(Issue: September 1999)

IUPAC name:	**1-Methyl-2,4-dinitrobenzene** (2,4-dinitrotoluene; 2,4-DNT)	**1-Methyl-2.6-dinitrobenzene** (2,6-dinitrotoluene; 2,6-DNT)	**1-Methyl-2,4,6-trinitrobenzene** (2,4,6-trinitrotoluene; 2,4,6-TNT)
CAS No.:	121-14-2	606-20-2	118-96-7
Molecular formula:	$C_7H_6N_2O_4$	$C_7H_6N_2O_4$	$C_7H_5N_3O_6$
Molar mass:	182.1 g/mol	182.1 g/mol	227.1 g/mol

The MAK-Collection Part III: Air Monitoring Methods, Vol. 9. DFG, Deutsche Forschungsgemeinschaft
Copyright © 2005 WILEY-VCH Verlag GmbH & Co. KGaA, Weinheim
ISBN: 3-527-31138-6

Sampling with a pump and adsorption on a combined sample holder consisting of a quartz fibre filter and a Tenax adsorption tube, HPLC after elution

Summary

This method permits the determination of concentrations of 2,4-dinitrotoluene (2,4-DNT), 2,6-dinitrotoluene (2,6-DNT) and 2,4,6-trinitrotoluene (2,4,6-TNT) in working areas, averaged over the sampling time after personal or stationary sampling.

Principle:	A pump is used to draw a measured volume of air through the combined sample holder, taking into account the definition of inhalable dust fraction [1]. In the process, the particulate portions are deposited on a quartz filter and the vapour portions are adsorbed onto Tenax. The portions on the filter are eluted with methanol directly after sampling. The sampling and backup sections are also covered with methanol when the Tenax tube is processed during sample preparation. After the compounds are eluted by treatment in an ultrasonic bath, 2,4-DNT, 2,6-DNT and 2,4,6-TNT are determined by liquid chromatography.

Technical data

Limit of quantification:	absolute: 1.5 ng for 2,4-DNT and 2,6-DNT; 1 ng for 2,4,6-TNT
	relative: 0.005 mg/m^3 for 2,4-DNT and 2,6-DNT for a 120-litre air sample, 2 mL elution solution and 5 μL injection volume; 0.007 mg/m^3 for 2,4,6-TNT for a 120-litre air sample, 4 mL elution solution and 5 μL injection volume
Selectivity:	The method is highly selective; it allows separation from other nitro compounds with good resolution (see Section 5.3). Values measured may be too high in the presence of interfering components. Interference can be identified by using a diode array detector and comparing spectra. Generally, interference can be eliminated by selecting different separation conditions.
Advantages:	Personal sampling and selective determinations possible
Disadvantages:	No indication of peak concentrations
Apparatus:	Pump Gas meter or volumetric flow meter Quartz fibre filter and Tenax tube with a holder Liquid chromatograph equipped with a UV detector

Detailed description of the method

Contents

1 Equipment, chemicals and solutions

1.1 Equipment

For sampling:
- Pump, suitable for flow rates of 1 L/min, e.g. Alpha 1 from Du Pont; supplier in Germany: DEHA Haan + Wittmer GmbH, D-71292 Friolzheim
- Gas meter or volumetric flow meter
- GGP-U 1.0 sampling head, e.g. from GSM, D-41469 Neuss
- Quartz fibre filter, diameter 37 mm, e.g. Ederol from Binzer & Munktell, D-35088 Battenberg
- Adsorption tube filled with Tenax ((standardised, consisting of two Tenax sections of about 100 mg (sampling section) and 50 mg (backup section) separated by glass wool)), e.g. Orbo 402 from Supelco, catalogue No. 20832, supplier in Germany: Sigma-Aldrich Chemie GmbH, Supelco division, D-82024 Taufkirchen
- Hygrometer

For sample preparation and analysis:
- Volumetric flasks, 2 mL
- Volumetric flasks, 500 mL and 1000 mL
- Adjustable piston pipettes, 10 μL – 5000 μL, e.g. from Eppendorf AG, D-22339 Hamburg

- Microlitre syringes, 1 μL, 10 μL, 25 μL, 50 μL, 100 μL, e.g. from Hamilton GmbH, D-64220 Darmstadt
- Screw-cap glass vials equipped with screw caps and PTFE-coated silicone septa, 5 mL and 10 mL
- Clear glass sample vials with screw caps and PTFE-coated silicone septa for HPLC with an autosampler, 1.5 mL
- Syringe filters for the filtration of organic samples, pore size 0.5 μm, e.g. Millex-FH$_{13}$, from Millipore, D-65824 Schwalbach
- Disposable syringes, 2 mL, 5 mL
- Disposable needles
- Wire hook for removing the inserts from the adsorption tubes, e.g. the Puller/Inserter tool, from Supelco, catalogue No. 22406
- Ultrapure water system for the preparation of ultrapure water, e.g. NANOpure ultrapure water system from Barnstead, supplier in Germany: Wilhelm Werner GmbH, D-51381 Leverkusen
- Ultrasonic bath
- Liquid chromatograph with a column thermostat and UV detector (variable wavelength)
- Data system

1.2 Chemicals and Solutions

2,4-Dinitrotoluene: Standard solution of 2000 μg/mL in methanol, e.g. from Wasagchemie Sythen GmbH, D-45721 Haltern

2,6-Dinitrotoluene: Standard solution of 2000 μg/mL in methanol, e.g. from Wasagchemie Sythen GmbH

2,4,6-Trinitrotoluene:
Standard solution of 2000 μg/mL in methanol, e.g. from Wasagchemie Sythen GmbH

The standard solutions must be kept in a refrigerator at + 4 °C and can be stored for at least six months.

For HPLC:
Ultrapure water, e.g. prepared with the ultrapure water system (UHQ water)
Methanol, HPLC Gradient Grade, e.g. from Baker, D-64347 Griesheim

Elution agent:	Methanol
Eluent:	300 mL methanol mixed with 700 mL UHQ water
Stock solution:	Solution of 25 μg/mL 2,4-DNT, 25 μg/mL 2,6-DNT and 15 μg/mL 2,4,6-TNT in methanol
	25 μL each of the 2,4-DNT and 2,6-DNT standard solutions and 15 μL of the 2,4,6-TNT standard solution are pipetted into a 2 mL volumetric flask containing approx. 1 mL methanol. The flask is then filled to the mark with methanol and shaken.
	The stock solution is stable for at least three months in the dark at room temperature.

Calibration
solutions: Solutions of 2,4-DNT and 2,6-DNT at 0.25 μg/mL to 2.5 μg/mL and
                              2,4,6-TNT at 0.15 μg/mL to 1.5 μg/mL in methanol (see Table 1).
                              Each calibration solution is prepared by first placing the solvent,
                              methanol, in a sample vial and then adding the appropriate volume
                              of stock solution with a pipette. The sample vials are sealed and
                              shaken.
                              The volumes of methanol and stock solution for preparation of the
                              calibration solutions can be taken from Table 1.

Table 1.

Solution	$V_{stock\ solution}$ μL	$V_{methanol}$ μL	Concentration μg/mL		
			2,4-DNT	2,6-DNT	TNT
1	10	990	0.25	0.25	0.15
2	20	980	0.50	0.50	0.30
3	30	970	0.75	0.75	0.45
4	40	960	1.00	1.00	0.60
5	50	950	1.25	1.25	0.75
6	60	940	1.50	1.50	0.90
7	70	930	1.75	1.75	1.05
8	80	920	2.00	2.00	1.20
9	90	910	2.25	2.25	1.35
10	100	900	2.50	2.50	1.50

                              For an air sample volume of 120 L, these solutions cover the fol-
                              lowing concentration ranges:
                              2,4-DNT/2,6-DNT: 0.005 mg/m^3 to 0.042 mg/m^3 for an elution vo-
                              lume of 2 mL and an injection volume of 5 μL.
                              2,4,6-TNT: 0.007 mg/m^3 to 0.05 mg/m^3 for an elution volume of 4
                              mL and an injection volume of 5 μL.
                              The calibration solutions are stable at room temperature for at least
                              three days.

2 Sampling

For sampling, the sampling head is fitted with the quartz fibre filter and the Tenax
tube. The pump and sampling head are carried by a person during working hours or
used in a stationary position.
At room temperature, 2,4,6-TNT occurs mainly in particulate form. To take into account
the definition of inhalable dust fraction [1] during sampling, the flow rate is set at 1 L/
min. With sampling for two hours this corresponds to an air sample volume of 120 litres.

In order to prevent low recoveries due to evaporation loss, the loaded quartz fibre filter should be processed immediately after sampling. The filter is placed in a screw-cap glass vial containing 4 mL methanol, and the vial is then sealed and shaken briefly. The Tenax tube should be closed tightly with the caps.

It is mandatory to measure air humidity, as recovery rates for 2,4,6-TNT depend on relative humidity (see Section 5.1).

3 Analytical determination

3.1 Sample preparation and analysis

For sample preparation, the sampling and backup sections of the Tenax tube are placed in separate screw-cap glass vials, covered with 2 mL methanol and treated for 15 minutes in an ultrasonic bath until elution is complete. The quartz fibre filter, treated with 4 mL methanol immediately after sampling (see Section 2), is also sonicated in the ultrasonic bath for 15 minutes. The vials are removed from the ultrasonic bath and allowed to cool to room temperature for about 30 minutes. Portions of liquid are taken from the supernatant solutions of the processed samples by means of disposable syringes and filtered into separate sample vials through syringe filters. The sample vials are sealed, 5 µL aliquots injected into the liquid chromatograph and chromatograms recorded under the operating conditions stated in Section 3.2.

To ensure that the methanol used for elution, the quartz fibre filter and the Tenax tube contain no interfering impurities, an unloaded quartz fibre filter and a Tenax tube are treated and analysed analogously (blank values). The quantitative evaluation of the chromatograms is performed by the external standard method.

3.2 Operating conditions for HPLC

The method was characterised under the following experimental conditions:

Apparatus:	Liquid chromatograph of the Merck-Hitachi LaChrom series, equipped with an L-7100 low-pressure gradient pump, L-7200 autosampler, L-7400 UV and L-7450 diode array detector from Merck, D-64293 Darmstadt
	Column block heater with temperature control, model 7970, from Jones Chromatography Ltd., supplier: VDS optilab, D-10365 Berlin
	Degasys DG-1310 degasser from VDS optilab
Column:	Stainless steel columns (length: 250 mm, internal diameter: 2 mm), filled with Hypersil ODS (particle size: 3 µm) from VDS optilab
Elution:	Isocratic elution

Eluent:	Mixture consisting of 30% (v/v) methanol and 70% (v/v) UHQ water
Flow rate:	0.2 mL/min
Injection volume:	5 μL, with injection loop
Detection wavelengths:	Initially 230 nm for 2,4,6-TNT and then 203 nm for the dinitrotoluenes
System pressure:	Approx. 240 bar
Furnace temperature:	40 °C

4 Calculations

4.1 Calibration

Aliquots of 2 μL of each of the calibration solutions exemplarily described in Section 1.2 are injected into the chromatograph and chromatograms are recorded. The peak areas determined are plotted against the corresponding concentrations of 2,4-DNT, 2,6-DNT and 2,4,6-TNT contained in the calibration solutions in order to construct the calibration curves. They are linear in the specified concentration range from Table 1.

4.2 Calculation of the analytical result

The concentrations by weight of 2,4-DNT, 2,6-DNT and 2,4,6-TNT in the air sample are calculated in mg/m^3 according to Equation (1):

$$c_w = \frac{c_x \cdot V_e}{V_l \cdot \eta} \tag{1}$$

where:

c_w is the concentration by weight of 2,4-DNT, 2,6-DNT or 2,4,6-TNT in the air sample, given in mg/m^3

c_x is the concentration of 2,4-DNT, 2,6-DNT or 2,4,6-TNT in μg/mL taken from the respective calibration curve

V_e is the elution volume in mL

V_l is the volume of the air sample in litres

η is the recovery.

When one of the components mentioned above is detected on both the quartz fibre filter and the Tenax (sampling section), the two results for that compound need to be added. If the amounts of analytes found on the backup section exceed by more than 10% of those deposited on the sampling section, sampling must be repeated with a smaller air sample volume.

5 Reliability of the method

The method was developed in accordance with DIN EN 482 [2].

5.1 Accuracy and recovery

The individual accuracies and recoveries in the minimum range of measurement were determined for four different concentrations (see Table 2). The quartz fibre filters were spiked directly with volumes of the appropriate standard solution as given in Table 2. Then laboratory air (30–40% relative humidity) was drawn through the sampling system (filter and Tenax tube) at a flow rate of 1 L/min for 2 hours (air sample volume 120 L). Each determination was carried out in replicates of 10.

The filters were processed as described in Section 2 immediately after simulated sampling. Sample preparation and analysis according to Section 3.1 yielded the values in Table 2.

Table 2.

Analyte	Weight	Volume of standard solution	Concentration standard deviation	Relative	Recovery
	μg	μL	mg/m^3	%	
2,4-DNT	0.6	0.3	0.005	3.7	0.89
	3.0	1.5	0.025	2.5	0.90
	6.0	3.0	0.050	2.1	0.91
	12.0	6.0	0.100	2.3	0.91
2,6-DNT	0.6	0.3	0.005	2.3	0.91
	3.0	1.5	0.025	3.2	0.93
	6.0	3.0	0.050	2.0	0.92
	12.0	6.0	0.100	2.4	0.91
2,4,6-TNT	1.2	0.6	0.010	3.4	0.92
	6.0	3.0	0.050	1.8	0.97
	12.0	6.0	0.100	2.5	0.97
	24.0	12.0	0.200	2.6	0.97

Mean recoveries for 2,4-DNT and 2,6-DNT were calculated from the recoveries for the individual concentration levels. The mean recoveries are given in Table 3 and should be used to correct the analytical results.

Table 3.

Analyte	Mean recovery
2,4-DNT	0.90
2,6-DNT	0.92

When determining 2,4,6-TNT content, it needs to be taken into account that the component's recovery depends on relative air humidity.

The effect of relative air humidity on sampling was investigated by spiking the filters for each of six sample holders with 6 μg 2,4-DNT, 6 μg 2,6-DNT and 12 μg 2,4,6-TNT. Subsequently, sampling was simulated for two hours at room temperature in a dynamic test gas apparatus at relative humidities of 40, 50, 60, 70 and 80%. The sample holders were processed and analysed according to Sections 2 and 3.1.

Whereas recoveries for DNT isomers did not exhibit dependence on relative air humidity, 2,4,6-TNT recovery steadily decreased with increasing humidity. Dependence was found to be linear in the range studied. No correction is necessary for relative humidities below 40%. The dependence of 2,4,6-TNT recovery on relative air humidity is shown in Figure 1.

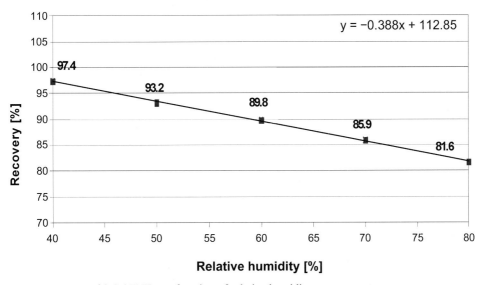

Fig. 1. Recovery of 2,4,6-TNT as a function of relative humidity.

5.2 Limit of quantification

The absolute limits of quantification for 2,4-DNT, 2,4-DNT and 2,4,6-TNT in methanol are 1.5 ng, 1.5 ng and 1 ng, respectively. They were determined using the linear calibration method according to DIN 32645 [3].

Calculation of the relative limits of quantification are based on the assumption that generally 2,4,6-TNT is present in the particulate form and is deposited on the filter, whereas the DNT isomers occur mainly in the vapour form and hence are deposited in the Tenax tube.

The relative limit of quantification is 0.005 mg/m^3 for 2,4-DNT and 2,6-DNT for a 120-litre air sample, 2 mL elution solution and an injection volume of 5 µL. For 2,4,6-TNT, the relative limit of quantification is found to be 0.007 mg/m^3 for a 120-litre air sample, 4 ml elution solution and 5 µL injection volume.

5.3 Selectivity

The selectivity of the method depends above all on the type of column and elution conditions selected. The separation conditions described here have proved reliable in practice. The following substances were also analysed under the operating conditions for liquid chromatography described in Section 3.2. They do not interfere with the determination of 2,4-DNT, 2,6-DNT or 2,4,6-TNT:

- Glycol dinitrate (nitroglycol)
- Diethylene glycol dinitrate
- Glycerol trinitrate (nitroglycerol, nitroglycerin)
- Nitrobenzene
- 1,3-Dinitrobenzene
- 1,4-Dinitrobenzene
- 1,3,5-Trinitrobenzene
- 1,3,5,7-Tetranitro-1,3,5,7-tetraza-cyclooctane (octogen, HMX)
- 1,3,5-Trinitro-1,3,5-triaza-cyclohexane (hexogen, RDX)
- 2,4,6-Trinitrophenol (picric acid)
- N-Methyl-2,4,6-N-tetranitroaniline (tetryl)
- 2-Amino-4-nitrotoluene
- 2-Amino-6-nitrotoluene
- 4-Amino-2,6-dinitrotoluene

Any interference from other substances can be identified by using a diode array detector and comparing spectra.

6 Discussion

It is essential to ensure that the quartz fibre filter is processed in 4 ml methanol immediately after sampling. The shelf-life of this solution and the loaded Tenax tubes is at least 16 days at room temperature without any loss of analyte.

7 References

[1] *Deutsches Institut für Normung e.V.* (DIN) (1993) DIN EN 481 – Arbeitsplatzatmosphäre – Festlegung der Teilchengrößenverteilung zur Messung luftgetragener Partikel. Beuth-Verlag, Berlin

[2] *Deutsches Institut für Normung e.V.* (DIN) (1994) DIN EN 482 – Arbeitsplatzatmosphäre – Allgemeine Anforderungen an Verfahren für Messung von chemischen Arbeitsstoffen. Beuth-Verlag, Berlin

[3] *Deutsches Institut für Normung e.V.* (DIN) (1994) DIN 32645 – Chemische Analytik – Nachweis-, Erfassungs- und Bestimmungsgrenze. Beuth Verlag, Berlin

Author: *D. Breuer*

Hydrogen fluoride and fluorides

Method number   1

Application   Air analysis

Analytical principle   Ion chromatography

Completed in   June 2004

Summary

The method permits the determination of particulate fluorides and hydrogen fluoride in a concentration range of 0.1 to 2 times the currently valid threshold limit value in air [1, 2]. Sampling is carried out by drawing ambient air through a double filter combination with a suitable sampling pump. The first, untreated filter collects the particulate fluorides while the second, sodium carbonate-impregnated filter traps the gaseous hydrogen fluoride. Elution is performed with an aqueous sodium carbonate/sodium hydrogen carbonate solution. Quantitative determination is carried out by ion chromatography.

Characteristics of the method

Accuracy:

Table 1. Standard deviation s_{rel} and mean variation u, $n = 10$ determinations for the determination of hydrogen fluoride.

Concentration mg/m^3	Standard deviation s_{rel} %	Mean variation u %
0.5	5.6	12.5
2.50	4.4	9.8
5.00	2.4	5.4

The MAK-Collection Part III: Air Monitoring Methods, Vol. 9. DFG, Deutsche Forschungsgemeinschaft
Copyright © 2005 WILEY-VCH Verlag GmbH & Co. KGaA, Weinheim
ISBN: 3-527-31138-6

Table 2. Standard deviation s_{rel} and mean variation u, $n = 10$ determinations for the determination of fluorides.

Concentration	Standard deviation	Mean variation
mg/m^3	s_{rel} %	u %
0.05	2.9	6.5
0.25	1.7	3.8
2.50	2.6	5.8
5.00	3.3	7.4

Limit of quantification:	0.33 mg/L $\triangleq$ 0.027 mg/m^3 fluoride for an air sample volume of 120 litres (corresponds to 0.03 mg hydrogen fluoride/m^3)
Recovery:	Fluorides: > 95 % Hydrogen fluoride: > 95 %
Sampling recommendation:	Sampling time: 2 hours Air sample volume: 120 litres Determination of short-term exposure limit possible

Hydrogen fluoride [CAS No. 7664-39-3]

Hydrogen fluoride (HF) is a colourless toxic gas (or liquid) with a pungent odour and is miscible with water in any proportion, yielding hydrofluoric acid. Its density at 0 °C is 1.002 g/cm^3 and its melting and boiling points are -83 °C and $+19.54$ °C, respectively.

HF and hydrofluoric acid are used in the manufacture of many fluorides that are of technological importance. Demand has declined in recent years because the use of chlorofluorocarbons (fluorochlorocarbons) as aerosol propellants has become strongly regulated and fluorine recycling in the aluminium industry has been improved. There have been new developments in the field of fluorine speciality chemicals with increasing use, for instance, in the following industries: fluorocarbons, fluoroplastics, fluorosurfactants, catalysts for condensation, isomerisation (e.g. in the production of branched-chain petroleum hydrocarbons) and polymerisation reactions. In the laboratory, HF is used as a non-aqueous solvent, in fluorination and in superacids [3].

The Deutsche Forschungsgemeinschaft proposes a threshold limit value of 1.7 mg/m^3 (2.0 mL/m^3) in air [1] and placement in peak limitation category I(1). In addition, HF was classed in Pregnancy Risk Group C.

According to the TRGS 900, the currently valid threshold limit value in air is 2.5 mg/m^3 with an excursion factor of =1= [2].

Fluorides

Fluorides are the salts of hydrofluoric acid. Replacement of the H atom in hydrogen fluoride by a metal yields simple, "neutral" fluorides such as sodium fluoride or calcium fluoride. The soluble fluorides are toxic as they inhibit acetylcholinesterase.

Fluorides are used as fluxing agents and wood preservatives and for special optical glasses and antireflection coatings on glass [4].

The currently valid MAK value for fluorides (calculated as fluorine) is 2.5 mg/m^3 – E – in the inhalable aerosol fraction. For the limitation of exposure peaks, fluorides have been classed in peak limitation category II(2) in the List of MAK and BAT Values [1]. According to the TRGS 900, the threshold value for air is 2.5 mg/m^3 with an excursion factor of 4 [2].

Authors: *D. Breuer, K. Gusbeth*
Examiners: *M. Tschickardt, M. R. Lahaniatis*

Hydrogen fluoride and fluorides

Method number   1

Application   Air analysis

Analytical principle   Ion chromatography

Completed in   June 2004

Contents

1 General principles

The method permits the determination of particulate fluorides and hydrogen fluoride in a concentration range of 0.1 to 2 times the currently valid threshold limit value in air [1, 2]. Sampling is carried out by drawing ambient air through a double filter combination with a suitable sampling pump. The first, untreated filter collects the particulate fluorides while the second, sodium carbonate-impregnated filter traps the gaseous hydrogen fluoride. Elution is performed with an aqueous sodium carbonate/sodium hydrogen carbonate solution. Quantitative determination is carried out by ion chromatography.

2 Equipment, chemicals and solutions

2.1 Equipment

Pump for personal air sampling, flow rate 1.0 L/min
Aerosol monitor for sampling inhalable aerosols, e.g. GSP with sampling head, from GSM GmbH, Neuss
Cellulose nitrate filter, diameter 37 mm, pore size 0.8 μm, e.g. from Sartorius AG, Göttingen
Polypropylene separating mesh, diameter 37 mm, approx. 1 mm mesh, <1 mm thick
Polyethylene bottles with screw tops, volume approx. 20 mL, e.g. from Müller-Ratiolab
10 to 1000 mL Polypropylene volumetric flasks
10–5000 μL Piston pipettes (adjustable)
Polypropylene (PP) autosampler vials, V = 0,75 mL, e.g. from Supelco, Sigma-Aldrich Chemie GmbH, Taufkirchen
Disposable filters for the filtration of aqueous samples, diameter 25 mm, pore size 0.45 μm, suitable for ion chromatography
Ultrasonic bath
Ultrapure water system
HPLC system with conductivity detector, e.g. from Dionex GmbH, Idstein

2.2 Chemicals

Fluoride standard solution: w (F$^-$) = 1000 mg/L, e.g. from Merck, Darmstadt
IC Instrument check standard: w (F$^-$) = 100 mg/L, w (Cl$^-$) = 200 mg/L, w (Br$^-$, NO^{3-}; SO$_4^{2-}$) = 400 mg/L, w (HPO$_4^{2-}$) = 600 mg/L, e.g. from SPEX Industries GmbH, Grasbrunn
The IC Instrument check standard is a control sample based on a certified standard.
Sodium carbonate anhydrous, analytical grade, e.g. from Fluka, Sigma-Aldrich Chemie GmbH, Taufkirchen
Sodium carbonate, analytical grade, e.g. from Fluka, Sigma-Aldrich Chemie GmbH, Taufkirchen

2.3 Solutions

Filter impregnation solution (c (Na$_2$CO$_3$) = 0.75 mol/L):
7.95 g sodium carbonate is weighed into a 100 mL volumetric flask, dissolved in ultra-pure water, and the flask is filled to the mark.

Eluent stock solution (0.8 mol/L Na$_2$CO$_3$, 0.1 mol/L NaHCO$_3$):
16.96 g sodium carbonate and 1.68 g sodium hydrogen carbonate are weighed into a 200 mL volumetric flask, dissolved in ultrapure water, and the flask is filled to the mark.

Eluent (8.0 mmol/L Na$_2$CO$_3$, 1.0 mmol/L NaHCO$_3$):
Ultrapure water is placed in a 1-litre volumetric flask; 10 mL eluent stock solution is added and the flask is filled to the mark with ultrapure water. As a rule, the eluent is prepared freshly only as needed.
After equipment idle times of more than seven days, the old eluent must be discarded and new eluent freshly prepared.

Eluent-rinsing solution (4.0 mmol/L Na$_2$CO$_3$, 0.5 mmol/L NaHCO$_3$):
Ultrapure water is placed in a 1-litre volumetric flask; 10 mL eluent stock solution is added and the flask is filled to the mark with ultrapure water.
As a rule, the rinsing solution is prepared freshly only as needed.

2.4 Calibration standards

Stock solution (100 mg/L fluoride):
2.5 mL fluoride standard solution is pipetted into a 25 mL volumetric flask, and the flask is filled to the mark with eluent. The solution is stable at room temperature for at least 4 weeks.

Table 3. Pipetting scheme for calibration standards for a concentration range of 0.4–4.0 µg/mL.

Calibration solution No.	Volume of the stock solution	Final volume of the calibration solution	Concentration of the calibration solution (F$^-$)	Concentration calculated as HF
	µL	mL	µg/L	mg/L
1	100	25	0.4	0.42
2	200	25	0.8	0.84
3	300	25	1.2	1.26
4	400	25	1.6	1.68
5	500	25	2.0	2.11
6	600	25	2.4	2.53
7	700	25	2.8	2.95
8	800	25	3.2	3.37
9	900	25	3.6	3.79
10	1000	25	4.0	4.21

The calibration solutions (Table 3) are prepared with ultrapure water, transferred to auto-sampler vials and analysed. They are stable at room temperature for at least two weeks.

3 Sample collection and preparation

3.1 Pretreatment of the sample carrier

The sample carrier consists of a combination of two serially arranged cellulose nitrate filters.
The first cellulose nitrate filter is used untreated. It collects the airborne fluorides.
The second filter is impregnated with 120 µL of impregnation solution (Section 2.3). The impregnation solution is evenly distributed on the filter using a pipette. Subsequently, the filter must be dried at room temperature for several hours.

3.2 Sampling and sample preparation

The dried impregnated filter is placed in the filter cassette, followed by the polypropylene separating mesh (Section 2.1) and finally the untreated filter. The filter cassette is assembled and sealed with the caps provided.
Care must be taken that the filter cassette is inserted into the GSP dust collection head with the flattened side down [5]. Figure 1 illustrates the design of the GSP dust collection head.
The dust collector is assembled and connected to a flow-regulated sampling pump. The pump is used to draw the sample air through the sampling system at a flow rate of 1.0 L/min. After sampling, the filter cassette is removed from the sampling system and sealed with the caps provided.
For analysis, the two cellulose nitrate filters are removed, placed in separate screw-cap bottles and covered with 10 mL elution solution. The sample containers are closed, treated for 15 minutes in an ultrasonic bath and allowed to stand for 30 minutes. Liquid from the supernatant solution of the prepared sample is drawn into a disposable syringe and filtered through a membrane filter into an autosampler vial. Analysis is performed by ion chromatography.

4 Operating conditions for ion chromatography

Apparatus:	GP 40 low-pressure gradient pump, ED 40 electrochemical detector, LC 10 liquid chromatography module, DX 500 series, AS 3500 autosampler with column oven, data system, from Dionex GmbH, Idstein

Degassing: Degasys DG 1310, VDS Optilab
Columns: Precolumn: IonPac AG14A (4 × 50 mm), from Dionex GmbH
                     Column: IonPac AS14A (4 × 250 mm), from Dionex GmbH
Suppressor: ASRS Ultra Anion Self-Regenerating Suppressor, 4 mm, from
                     Dionex GmbH
Eluent: 8.0 mmol/L Na_2CO_3; 1.0 mmol/L $NaHCO_3$
Eluent flow rate: 1.0 mL/min
Injection volume: 50 μL
Column temperature: 30 °C
System pressure: approx. 13.0 MPa (approx. 130 bar)

Figure 2 shows a chromatogram obtained under the conditions given above.

5 Analytical determination

A 50 μL aliquot of the prepared sample solution is injected via the autosampler and analysed under the conditions given above.

If the analytical results are not within the calibration range, the samples must be diluted appropriately and analysed again.

6 Calibration

A calibration curve is constructed using the calibration solutions described in Section 2.4. Volumes of 50 μL of each of the calibration solutions are injected and analysed as for the sample solutions. The peak areas obtained are plotted against the corresponding concentrations. The calibration curve for fluoride is linear in the range from 0.4 mg/L to 4.0 mg/L (hydrogen fluoride 0.42 to 4.21 mg/L). Figure 3 shows an example of a fluoride calibration curve.

To check the calibration function, a control sample should be analysed each day. A new calibration curve must be constructed if the analytical conditions change or quality control shows this to be necessary.

7 Calculation of the analytical result

The concentration of hydrogen fluoride and other fluorides in the workplace air is calculated using the fluoride concentrations in the sample solutions as calculated by the data system.

Based on the concentration in the sample solution, the concentrations of hydrogen fluoride and other fluorides in the workplace air are calculated taking into account dilution steps and the air sample volume.

The peak area of the sample signal is calculated according to the following equation:

$$F_{corr} = F - F_{blank} \tag{1}$$

where:
F is the peak area from the sample chromatogram
F_{korr} is the peak area after correction for the blank value
F_{blank} is the peak area for the blank value

The following equations apply for calculation of the concentration of hydrogen fluoride in the workplace air:

$$\rho = \frac{(F_{corr} - a)}{b \cdot V_{air} \cdot \eta} \cdot f_{HF} \cdot 0.01\, l \cdot \frac{273 + t_g}{273 + t_a} \; [mg/m^3] \tag{2}$$

The same equation can be used to calculate the analytical result for fluorides, except that this requires no conversion factor "f".
At 20 °C and 1013 hPa:

$$\rho_O = \rho \frac{273 + t_a}{293} \cdot \frac{1013}{p_a} \; [mg/m^3] \tag{3}$$

The corresponding concentration by volume σ – independently of the state variables pressure and temperature – is given by:

$$\sigma = \rho_O \cdot \frac{V_m}{M} \tag{4}$$

$$\sigma = \rho \frac{273 + t_a}{p_a} \cdot \frac{1013}{293} \cdot \frac{V_m}{M} \tag{5}$$

For hydrogen fluoride at $t_a = 20\,°C$ and $p_a = 1013\,hPa$:

$$\sigma\,(HF) = \rho \cdot 1.202 \, \frac{mL}{m^3} \tag{6}$$

where:
ρ is the concentration by weight in the ambient air as a function of t_a and p_a
ρ_o is the concentration by weight in the ambient air at 20 °C and 1013 hPa
a is the intercept of the calibration curve
b is the slope of the calibration curve
η is the recovery (to be taken into account where appropriate)
f_{HF} is the stoichiometric conversion factor $F^- \rightarrow HF$ (1.053)
$0.01\,l$ is the conversion factor for the volume of the measured sample
V_{air} is the air sample volume in m^3
t_g is the temperature in the gasmeter in °C
t_a is the temperature of the ambient air in °C

p_a is the atmospheric pressure of the ambient air in hPa
σ is the HF concentration in the ambient air in mL/m^3
V_m is the molar volume of hydrogen fluoride in L/mol
M is the molar mass of hydrogen fluoride in g/mol

8 Reliability of the method

The characteristics of the method were determined according to the standard EN 482 [6]. The limit of detection and limit of quantification were determined according to DIN 32645 [7].

A dynamic test gas apparatus (relative humidity ~50%) was used to determine the precision in the minimum measurement range (0.17–3.4 mg/m^3) during the sampling of hydrogen fluoride. The characteristics for fluorides were determined by spiking filters with standard solutions. Sample preparation was carried out as described in Section 3.2.

8.1 Precision

Six samples were loaded per concentration (Table 4 and 5).

Table 4. Standard deviation s_{rel} and mean variation u, $n = 10$ determinations for the determination of hydrogen fluoride.

Concentration	Standard deviation	Mean variation
	s_{rel}	u
mg/m^3	%	%
0.5	5.6	12.5
2.50	4.4	9.8
5.00	2.4	5.4

Table 5. Standard deviation s_{rel} and mean variation u, $n = 10$ determinations for the determination of fluorides.

Concentration	Standard deviation	Mean variation
	s_{rel}	u
mg/m^3	%	%
0.05	2.9	6.5
0.25	1.7	3.8
2.50	2.6	5.8
5.00	3.3	7.4

8.2 Recovery

The recovery was checked for the whole calibration range. This was accomplished using the sample carriers prepared for the determination of precision in the minimum measurement range. The results were compared with the standard solutions used for ca-libration.

No significant losses were noted for hydrogen fluoride or other fluorides, with mean re-coveries exceeding 95% ($\eta > 0.95$) for both HF and F⁻.

8.3 Limit of quantification

The limit of quantification for fluoride was calculated according to DIN 32645. It is 0.027 mg/m^3 (HF 0.03 mg/m^3) for an air sample volume of 120 litres, 10 mL sample solution and a 50 µL injection volume.

8.4 Shelf-life

Hydrogen fluoride:
To determine the shelf-life, two series of impregnated filters were loaded with known concentrations of hydrogen fluoride (low and high concentration) using a dynamic test gas apparatus. The filters were loaded and analysed under the sampling conditions de-scribed in Sections 3.2 and 5.

The loaded filters were initially stored at room temperature for seven days and then kept in a refrigerator at +4 °C. Duplicate determinations were carried out every week. No significant changes in hydrogen fluoride concentration were observed over a period of 4 weeks.

Fluorides:
To determine the shelf-life, two series of cellulose nitrate filters were loaded using a microlitre syringe (low and high concentration). Initial storage for 7 days was at room temperature, followed by storage in the refrigerator at +4 °C. Duplicate determinations were carried out every week.

No change in the fluoride concentration was noted over a period of 2 weeks (recovery >95%). At the low concentration, the recovery decreased to about 93% over the fol-lowing weeks.

Loaded sample carriers should be analysed within 14 days.

8.5 Blank value

The cellulose nitrate filters, and particularly the impregnated blank filters, have a de-tectable blank value which varies from batch to batch. The blank value must therefore be checked for each batch and taken into account when calculating the results.

9 Discussion of the method

As a result of the limited shelf-life, analysis of loaded sample carriers for particulate fluorides must be carried out within 14 days.
Humidity had no notable influence in the range from 30 to 70% relative humidity.

10 References

[1] *Deutsche Forschungsgemeinschaft* (2004) List of MAK und BAT Values 2004. Commission for the Investigation of Health Hazards of Chemical Compounds in the Work Area. Report No. 40. Wiley-VCH Verlag, Weinheim.

[2] Bundesministerium für Arbeit und Sozialordnung (2000) TRGS 900: Grenzwerte in der Luft am Arbeitsplatz-Luftgrenzwerte. Technische Regeln und Richtlinien des BMA zur Verordnung über gefährliche Stoffe. Ausgabe Oktober 2000, BArbBl. 10/2000, S. 34–63, letzte Änderung 5/2004.

[3] Römpp-Online – Römpp-Lexikon der Chemie, Fluorwasserstoff (2003), 9. Auflage, Georg Thieme Verlag, Stuttgart, New York.

[4] Römpp-Online – Römpp-Lexikon der Chemie, Fluoride (2003), 9. Auflage, Georg Thieme Verlag, Stuttgart, New York.

[5] Geräte zur Probenahme der einatembaren Staubfraktion (E-Staub) (Kennzahl 3010). In: BIA-Arbeitsmappe Messung von Gefahrstoffen. Hrsg.: Berufsgenossenschaftliches Institut für Arbeitssicherheit – BIA, Sankt Augustin. Bielefeld: Erich Schmidt – Losebl.-Ausg.

[6] *European Committee for Standardization* (CEN) (1994) EN 482 – Workplaceatmospheres – General requirements for the performance of procedures for the measurement of chemical agents. Brussels.

[7] *Deutsches Institut für Normung e.V.* (DIN) (1994) DIN 32645– Chemische Analytik-Nachweis-, Erfassungs- und Bestimmungsgrenze, Beuth Verlag, Berlin

Authors: *D. Breuer, K. Gusbeth*
Examiners: *M. Tschickardt, M. R. Lahaniatis*

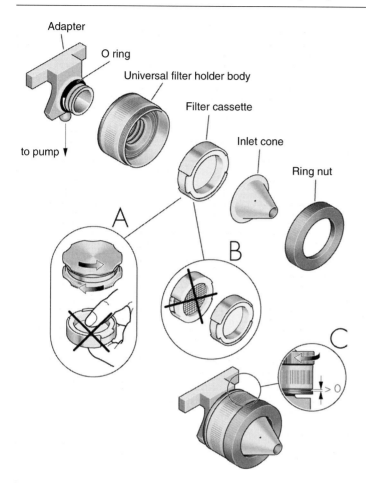

Adapter

O ring

Universal filter holder body

Filter cassette

Inlet cone

to pump ▼

Ring nut

A

B

C

Fig. 1. The GSP (total dust sampling) system developed by BIA, the Institute for Occupational Safety and Health of the German Federation of Institutions for Statutory Accident Insurance and Prevention (HVBG).

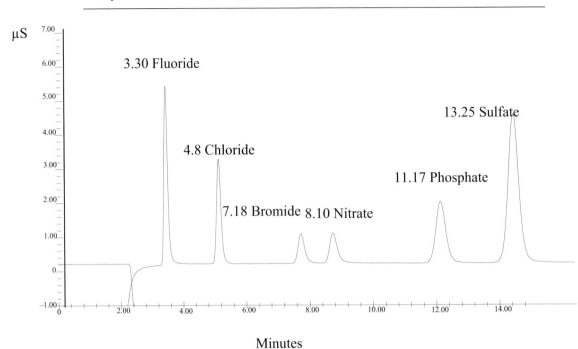

Fig. 2. Example of a chromatogram for the separation of the anions. w (fluoride) = 2.2 mg/L.

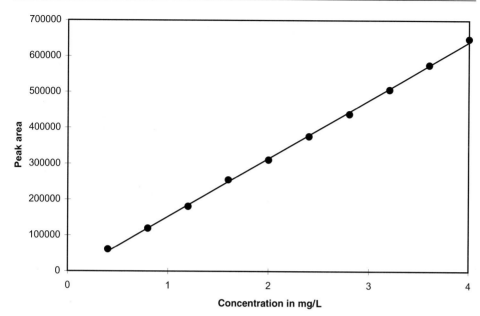

Fig. 3. Calibration curve for fluoride in the range of 0.4–4 mg/L.
♦ Measured value, —— Regression line

Halogenated anaesthetic gases (halothane, enflurane, isoflurane)

Method number	2
Application	Air analysis
Analytical principle	Gas chromatography/electron capture detector
Completed in	May 2003

Summary

The present method permits the simultaneous detection of the anaesthetic gases halothane (2-bromo-2-chloro-1,1,1-trifluoroethane), enflurane (2-chloro-1,1,2-trifluoroethyl difluoromethyl ether) and isoflurane (1-chloro-2,2,2-trifluoroethyl difluoromethyl ether) in the workplace air. Sampling occurs passively through diffusion of the halogenated anaesthetic gases from the ambient air onto the activated carbon contained in a glass tube, where they accumulate. Analysis after desorption with toluene is performed by gas chromatography using an electron capture detector (ECD). Quantitative evaluation is carried out using a calibration curve.

The MAK-Collection Part III: Air Monitoring Methods, Vol. 9. DFG, Deutsche Forschungsgemeinschaft
Copyright © 2005 WILEY-VCH Verlag GmbH & Co. KGaA, Weinheim
ISBN: 3-527-31138-6

Characteristics of the method

Precision:

Table 1. Standard deviation s_{rel} and mean variation u, $n = 6$ determinations.

Substance	Determinations n	Concentration mg/m^3	Standard deviation s_{rel} %	Mean variation u %
Halothane	6	8.9	4.8	12.3
	6	39.7	8.0	20.5
	5	80.72	3.7	9.6
Enflurane	6	32.8	4.8	12.3
	6	146.8	5.2	13.4
	5	298.12	4.0	10.3
Isoflurane	6	25.3	3.4	8.7
	6	113.0	8.0	20.6
	5	229.62	4.3	11.0

Limit of quantification: Halothane 0.03 mg/m^3
Enflurane 0.12 mg/m^3
Isoflurane 0.12 mg/m^3
with a sampling time of 8 hours

Recovery: Halothane $\eta = 0.95$ (95%)
Enflurane $\eta = 0.98$ (98%)
Isoflurane $\eta = 0.98$ (98%)

Sampling recommendation: Sampling time: 1–8 hours

Halogenated anaesthetic gases

Anaesthesia of patients undergoing medical procedures is performed, inter alia, with the volatile anaesthetics halothane, enflurane and isoflurane in combination with nitrous oxide.

Halothane (2-bromo-2-chloro-1,1,1-trifluoroethane) [CAS No. 151-67-7]

F$_3$C-CHBrCl

Halothane (molar mass 197.4 g/mol, boiling point +50.2 °C) is a reliable anaesthetic. It is known by several synonyms (fluothane, halan, narcotane and various further names). Concentration levels of 0.5–1.5% by volume in the inhaled air induce a state of general anaesthesia [1]. The main metabolite in warm-blooded animals is trifluoroacetic acid. The MAK value valid since 1978 is 41 mg/m^3, or 5 mL/m^3 (ppm), and the substance is currently under review by the Commission for the Investigation of Health Hazards of Chemical Compounds in the Work Area [2, 3]. The peak limitation for halothane has been assigned to category II(8) in the List of MAK and BAT Values [2, 3] and given an excursion factor of 4 according to the TRGS 900. Based on the information available, it cannot be precluded that halothane causes prenatal toxicity even if the MAK and BAT values are observed. Therefore, as a precaution, the substance has been assigned to Pregnancy risk group B.

Enflurane (2-chloro-1,1,2-trifluoroethyl difluoromethyl ether) [CAS No. 13838-16-9]

CHF$_2$-O-CF$_2$CHClF

Enflurane (molar mass 184.5 g/mol, boiling point +56.5 °C; synonym: ethrane) is a stable, non-flammable inhalation anaesthetic. It is mostly used in combination with nitrous oxide at a concentration level of 0.5–1.5% by volume [4]. Since 1994, the MAK value for enflurane has been set at 150 mg/m^3, or 20 mL/m^3 (ppm). The peak limitation for enflurane has been assigned to category II(8) in the List of MAK and BAT Values [2, 3] and given an excursion factor of 4 according to the TRGS 900.

Isoflurane (1-chloro-2,2,2-trifluoroethyl difluoromethyl ether) [CAS No. 26675-46-7]

CF$_3$-CHCl-O-CHF$_2$

Isoflurane (molar mass 184.5 g/mol, boiling point +48.5 °C) is a halogenated methyl ethyl ether with a central depressant effect and is used as an inhalation anaesthetic. It is also known by the synonyms forane, forene and aerrane. The induction and maintenance of general anaesthesia require an average inspiratory concentration of 2.3% by volume in the inhaled air or 1.2% by volume in combination with 50–70% nitrous oxide [5]. It is not possible to establish a MAK value on the basis of the insufficient data currently available. Therefore, isoflurane has been classed in Section IIb of the List of MAK and BAT Values. In default of national threshold limit values in air , the Hamburg 'Amt für Arbeitsschutz' (occupational health and safety office) in 1989 introduced a guidance value of 150 mg/m^3 for workplace exposure to isoflurane [6]. In the spring of 1997, the United Kingdom threshold limit value in air of 80 mg/m^3 was adopted into the TRGS 900 and is now equal in status with the German threshold limit values in air [3].

Author: *K.-H. Pannwitz*
Examiner: *M. Tschickardt*

Halogenated anaesthetic gases (halothane, enflurane, isoflurane)

Method number   2

Application   Air analysis

Analytical principle Gas chromatography/electron capture detector

Completed in   May 2003

Contents

1 General principles

The present method permits the simultaneous determination of the concentration levels in the workplace air of the halogenated anaesthetic gases halothane (2-bromo-2-chloro-1,1,1-trifluoroethane), enflurane (2-chloro-1,1,2-trifluoroethyl difluoromethyl ether) and isoflurane (1-chloro-2,2,2-trifluoroethyl difluoromethyl ether). Sampling occurs passively through diffusion of the halogenated anaesthetic gases from the ambient air onto the activated carbon contained in a glass tube, where they accumulate. Analysis after desorption with toluene is performed by gas chromatography/ECD. Quantitative evaluation is carried out using a calibration curve.

2 Equipment, chemicals and solutions

2.1 Equipment

Diffusion sampler with holder (ORSA, from Dräger Safety AG & Co. KGaA, Lübeck,
   Catalogue No. 6728 891) (see Figure 1)
Tweezers
15 mL sample vials with PTFE-lined screw caps
2 mL autosampler vials with PTFE-lined caps
Laboratory shaker
10, 25 and 100 mL volumetric flasks
2 and 10 mL graduated pipettes
5, 20, 100 and 200 μL microlitre syringes
Gas chromatograph with electron capture detector (ECD)
Computerised data collection and integration system

2.2 Chemicals

Toluene for residue analysis (e.g. LGC Promochem GmbH, Wesel)
Halothane 99.9%, prescription-only (supplied by pharmacies)
Enflurane 99.9%, prescription-only (supplied by pharmacies)
Isoflurane 99.9%, prescription-only (supplied by pharmacies)
Argon/methane 95/5%, ECD quality
Helium, 99.996%

2.3 Calibration standards

Calibration standards are to be prepared in the analyte concentration ranges from 0.1 to 2 times the valid threshold limit values in air, in accordance with EN 482 [8]. For a sampling time of 8 hours, the calibration standards should be prepared in the following concentration ranges (Table 2).

Table 2. Concentration ranges of the calibration standards.

Substance	Threshold limit value in air mg/m^3	Concentration range µg/mL	Density at 20 °C g/mL
Halothane	41	1.09–21.8	1.87
Enflurane	150	3.82–76.4	1.49
Isoflurane	80	2.03–40.7	1.52

Stock solution:
A stock solution is prepared which contains a mixture of the anaesthetic gases to be analysed. First, toluene is placed in a 25 mL volumetric flask and then 2 mL halothane, 10 mL enflurane and 5 mL isoflurane are added. Toluene is then added to the mark. The solution contains halothane, enflurane and isoflurane at 0.15 g/mL, 0.596 g/mL and 0.304 g/mL, respectively.

Working solution:
The working solution is prepared by pipetting 1 mL stock solution into a 100 mL volumetric flask containing 5 mL toluene. The flask is then filled to the mark with toluene.

Calibration standards:
Calibration standards are obtained by diluting the working solution with toluene according to the following pipetting scheme (Table 3):

Table 3. Pipetting scheme for the preparation of the calibration standards.

Volume of working solution µL	Final volume of calibration standard mL	Halothane concentration µg/mL	Enflurane concentration µg/mL	Isoflurane concentration µg/mL
5	10	0.75	2.98	1.52
10	10	1.50	5.96	3.04
20	10	2.99	11.92	6.08
50	10	7.48	29.8	15.2
100	10	14.96	59.6	30.4
150	10	22.44	89.4	45.6

A reduced sampling time requires appropriate dilution of the calibration standards.

3 Sampling

The ORSA diffusion sampler is removed from the transport container and mounted in its holder. Prepared in this manner, the sampling system is affixed to the clothes in the person's breathing zone. The openings of the diffusion sampler should not be obstructed or covered by textiles. The ambient conditions during sampling (temperature, atmospheric pressure, relative humidity) are determined and recorded in a sampling record together with the times when sampling was commenced and concluded. After sampling, the ORSA diffusion sampler is immediately placed in the transport container, which is then sealed. The duration of sampling should be between one and eight hours.

It is recommended to analyse loaded ORSA diffusion samplers within 2 weeks. The samplers must be kept in the refrigerator after transport to the laboratory.

4 Sample preparation

The ORSA diffusion sampler is removed from the transport container. A pair of tweezers is used carefully to remove one stopper, and the contents of the tube (activated carbon) are then transferred quantitatively to a sample vial. Subsequently, 10 mL of toluene is added with a pipette and the sample vial is sealed. Desorption is achieved by agitation on a laboratory shaker for one hour. After sedimentation of the activated carbon, a portion of the sample solution is removed and filled into an autosampler vial, which is then sealed with a PTFE-lined screw cap.

5 Operating conditions for gas chromatography

Apparatus:	Siemens Sichromat 2	
Column:	Material:	Fused silica
	Length:	30 m
	Internal diameter:	0.32 mm
	Stationary phase:	DB-1
	Film thickness:	1 μm
Detectors:	ECD	
Temperatures:	Furnace:	40 °C, 30 minutes isothermal
	Detector:	350 °C
	Injector:	200 °C
Carrier gas:	Helium, 5 mL/min	
Detector gas:	Argon/methane, 30 mL/min	
Injection volume:	1 μL	
Split:	40 mL/min	

Figure 2 shows an example of a chromatogram obtained under the conditions given above.

6 Calibration

Calibration is carried out under the same conditions as sample analysis. For this purpose, the calibration standards prepared as described in Section 2.3 are analysed under the operating conditions given in Section 5. The calibration curves for halothane, enflurane and isoflurane are constructed from the measured peak areas and the analyte concentrations of each calibration standard. It may be necessary to prepare further dilutions of the calibration standards in order to remain within the calibration range of the detector. Figure 3 shows an example of a calibration curve.

7 Calculation of the analytical result

The obtained peak areas and the corresponding calibration curve are used to calculate the weight of the respective halogenated anaesthetic gas contained in 10 mL. The concentration by weight ρ in mg/m^3 is calculated according to the following equation:

$$\rho = \frac{X \cdot 1000}{U \cdot \eta \cdot \tau_v} \tag{1}$$

At 20 °C and 1013 hPa:

$$\rho_0 = \rho \cdot \frac{273 + t}{293} \cdot \frac{1013\,\text{mbar}}{p} \tag{2}$$

The corresponding concentration by volume – independently of the state variables pressure and temperature – is given by:

$$\sigma = \rho \cdot \frac{273 + t_a}{293} \cdot \frac{1013}{p_a} \cdot \frac{V_m}{M} \tag{3}$$

At t_a = 20 °C and p_a = 1013 hPa:

$$\sigma\,(\text{halothane}) = \rho_{\text{halothane}} \cdot 0.122\,\frac{\text{mL}}{\text{mg}} \tag{4}$$

$$\sigma\,(\text{enflurane}) = \rho_{\text{enflurane}} \cdot 0.131\,\frac{\text{mL}}{\text{mg}} \tag{5}$$

$$\sigma \text{ (isoflurane)} = \rho_{\text{isoflurane}} \cdot 0.131 \; \frac{\text{mL}}{\text{mg}} \tag{6}$$

where:

X is the weight of the analyte in μg
U is the sampling rate of the halogenated anaesthetic gas in mL/min
    U (halothane): 5.70 mL/min
    U (enflurane): 5.31 mL/min
    U (isoflurane): 5.30 mL/min
η is the recovery
τ_v is the sampling time in minutes
t_a is the sampling site temperature in °C
p_a is the air pressure at the sampling site in hPa
ρ is the concentration by weight of the halogenated anaesthetic gas in the ambient air in mg/m^3 at t_a and p_a
ρ_o is the concentration by weight of the halogenated anaesthetic gas in the ambient air in mg/m^3 at 20 °C and 1013 hPa
M is the molar mass of the halogenated anaesthetic gas in g/mol
V_m is the molar volume in L/mol
σ is the concentration by volume of the halogenated anaesthetic gas in mL/m^3

8 Reliability of the method

8.1 Precision

To determine the precision of the method, concentrations between approx. 1/10 and twice the threshold limit value in air were prepared using a dynamic test atmosphere generation system [7, 8], and 6 ORSA diffusion samplers per test series were loaded for 4 hours. Sample preparation and analysis were performed according to Sections 4 and 5. The data shown in Table 4 were obtained.

Table 4. Standard deviation s_{rel} and mean variation u, $n = 5$ or 6 determinations.

Substance	Determinations	Concentration	Standard deviation	Mean variation
	n	mg/m^3	s_{rel} %	u %
Halothane	6	8.9	4.8	12.3
	6	39.7	8.0	20.5
	5	80.72	3.7	9.6
Enflurane	6	32.8	4.8	12.3
	6	146.8	5.2	13.4
	5	298.12	4.0	10.3
Isoflurane	6	25.3	3.4	8.7
	6	113.0	8.0	20.6
	5	229.62	4.3	11.0

8.2 Recovery

The recovery (desorption efficiency) was determined by analysis of ORSA diffusion samplers loaded with known concentrations in the range between 1/10 and twice the threshold limit value in air after a sampling time of 4 hours. The following values were determined:

Halothane: $\eta = 0.95$ (95%)
Enflurane: $\eta = 0.98$ (98%)
Isoflurane: $\eta = 0.98$ (98%)

8.3 Limit of quantification

The following limits of quantification were determined for a sampling time of 8 hours under the analytical conditions given in Section 5:

Halothane: 0.03 mg/m^3
Enflurane: 0.12 mg/m^3
Isoflurane: 0.12 mg/m^3

The ECD sensitivity depends on the detector settings. Sensitivity increases with temperature. The maximum detector temperature is limited by the type of column used in the apparatus.

8.4 Interference

Other hazardous substances that occur in operating theatre areas (anaesthetic room, operating theatre, recovery room), e. g. ethanol and 2-propanol, do not interfere with the analysis of halothane, enflurane or isoflurane under the analytical conditions given in Section 5.

9 Discussion of the method

The method is suitable for exposure measurements in hospital operating theatre areas for reasons of hygiene, in particular. The ORSA diffusion samplers are affixed to the clothes in the breathing zone; they to not interfere with work performance. During the testing of the method, the halogenated anaesthetic gases were determined in the concentration range between 1/10 and twice the valid threshold values in air over a sampling time of 4 hours. The sampling rate is independent of the sampling time. Measurements for monitoring short term exposure limit requirements are possible. In this context, particular attention must be paid to the exact determination of sampling time.

The column specified here has proved reliable. Retention times need to be checked when using other columns.

The present sampling method is, in principle, also suitable for desflurane and sevoflurane. Because ECD sensitivity for sevoflurane is low (1/1000 relative to halothane), analysis requires the use of a different desorption agent and detector.

10 References

[1] *Henschler D.* (1979) Halothan. In: H. Greim (Hrsg.) Gesundheitsschädliche Arbeitsstoffe. Toxikologisch-arbeitsmedizinische Begründungen von MAK-Werten. Band 1. Senatskommission zur Prüfung gesundheitsschädlicher Arbeitsstoffe VCH-Verlag, Weinheim.

[2] *Deutsche Forschungsgemeinschaft* (2004) List of MAK und BAT Values 2004. Commission for the Investigation of Health Hazards of Chemical Compounds in the Work Area. Report No. 40. WILEY-VCH-Verlagsgesellschaft, Weinheim.

[3] Technische Regel für Gefahrstoffe: Grenzwerte in der Luft am Arbeitsplatz – Luftgrenzwerte (TRGS 900). BArbBl. 10/2000 S. 34; 4/2001 S. 56; 9/2001 S. 86; 3/2002 S. 71; 3/2003 S. 69; 4/2003 S. 80, zuletzt geändert BArbBl. 4/2003 S. 80.

[4] *Greim H.* (1994) 2-Chlor-1,1,2-trifluorethyldifluormethylether (Enfluran). In: Gesundheitsschädliche Arbeitsstoffe. Toxikologisch-arbeitsmedizinische Begründungen von MAK-Werten. Band 1. Senatskommission zur Prüfung gesundheitsschädlicher Arbeitsstoffe Wiley-VCH Verlag, Weinheim.

[5] *Greim H.* (1993) Isofluran. In: Gesundheitsschädliche Arbeitsstoffe. Toxikologisch-arbeitsmedizinische Begründungen von MAK-Werten. Band 1. Senatskommission zur Prüfung gesundheitsschädlicher Arbeitsstoffe Wiley-VCH Verlag, Weinheim.

[6] *Freie und Hansestadt Hamburg, Behörde für Arbeit, Gesundheit und Soziales, Amt für Arbeitsschutz* (August 1991) Merkblatt für den Umgang mit Narkosegasen.

[7] *Verein Deutscher Ingenieure* (VDI) (1981) VDI-Richtlinie 3490 Blatt 8, Prüfgase-Herstellung durch kontinuierliche Injektion. Beuth Verlag, Berlin.

[8] *Europäisches Komitee für Normung* (CEN) (1994) DIN EN 482 – Arbeitsplatzatmosphäre – Allgemeine Anforderungen an Verfahren zur Messung von chemischen Arbeitsstoffen. Brüssel. Beuth Verlag, Berlin.

Author: *K.-H. Pannwitz*
Examiner: *M. Tschickardt*

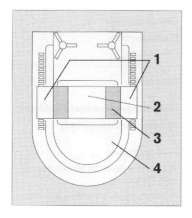

1 Diffusion path (cellulose acetate)
2 Recording area
3 Adsorption layer (activated carbon)
4 Holder

Fig. 1. ORSA diffusion sampler with holder.

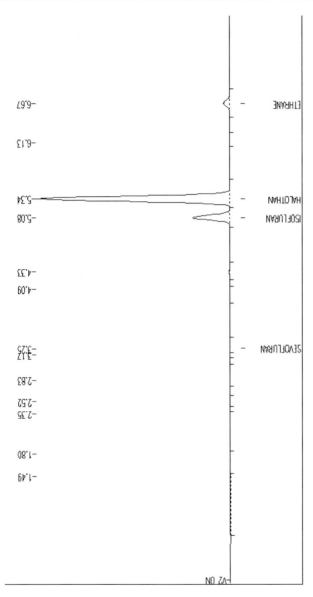

Fig. 2. Example of a calibration standard chromatogram of the halogenated anaesthetic gases (for analytical conditions see Section 5).

Isoflurane: 113 mg/m^3
Halothane: 39.7 mg/m^3
Enflurane: 147 mg/m^3

Isoflurane

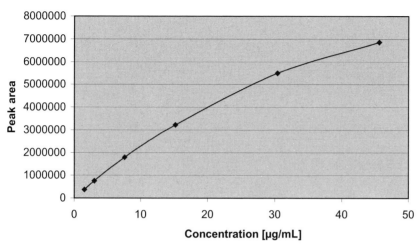

Fig. 3. Example of a calibration curve for isoflurane

Federation of the Employment Accidents Insurance Institutions of Germany
(Hauptverband der Berufsgenossenschaften)
Centre for Accident Prevention and Occupational Medicine
Alte Heerstraße 111, 53757 Sankt Augustin
Expert Committee Chemistry

Carcinogenic substances	Order number: BGI 505-68E
Established methods	Issue: August 2001

Method for the determination of sulfuric acid

Method tested and recommended by the Berufsgenossenschaften for the determination of sulfuric acid in working areas after discontinuous sampling.
Both personal and stationary sampling can be conducted for the assessment of working areas.

Sampling with a pump and collection on a filter, ion chromatography (HPIC) after elution.
Sulfuric acid-1-Ion chromatography
Sampling with quartz fibre filters
(Issue: August 2001)

Name:	Sulfuric acid
CAS No.:	7664-93-9
Molecular formula:	H_2SO_4
Molar mass:	98.07 g/mol

The MAK-Collection Part III: Air Monitoring Methods, Vol. 9. DFG, Deutsche Forschungsgemeinschaft
Copyright © 2005 WILEY-VCH Verlag GmbH & Co. KGaA, Weinheim
ISBN: 3-527-31138-6

Sampling with a pump and collection on a filter, ion chromatography after elution

Summary

This method permits the determination of sulfuric acid concentrations in working areas averaged over the sampling time after personal or stationary sampling.

Principle:	A pump is used to draw a measured volume of air through a quartz fibre filter. The collected sulfuric acid is eluted with a sodium carbonate/sodium hydrogen carbonate solution and determined by ion chromatography.

Technical data:

Limit of quantification:	absolute: 50 ng sulfuric acid relative: 0.01 mg/m^3 H$_2$SO$_4$ for a 420-litre air sample volume, 4 mL elution volume and 50 µL injection volume
Selectivity:	The selectivity of the chromatographic method depends above all on the type of column and the separation conditions used. The separation conditions described here have proved reliable in practice.
Advantages:	Personal sampling and selective determinations possible
Disadvantages:	No indication of peak concentrations
Apparatus:	Pump Gas meter or volumetric flow meter Binder-free quartz fibre filter and filter holder Ion chromatograph with conductivity detector

Detailed description of the method

Contents

1 Equipment, chemicals and solutions

1.1 Equipment

For sampling:
- Pump for personal air sampling, suitable for a flow rate of 210 L/hour, e. g. Gilian PP5-Ex, supplier in Germany: e. g. DEHA Haan + Wittmer GmbH, D-71292 Friolzheim
- Filter holder for sampling the inhalable dust fraction, e. g. GSP [1], from GSM GmbH, D-41469 Neuss
- Polyethylene bottles with screw caps, 16 × 56 mm, 6 mL
- Gas meter or volumetric flow meter
- Quartz fibre filter, binder-free, diameter 37 mm, e. g. Ederol T 293 from Binzer & Munktell, D-35088 Battenberg

For sample preparation and analysis:
- Volumetric flasks, 10 to 2000 mL
- Adjustable piston pipettes, 10 µL to 5000 µL, e. g. from Eppendorf AG, D-22339 Hamburg
- Disposable syringes, 5 mL with disposable needles (1.20 × 75 mm)
- Disposable filters for aqueous samples, diameter 25 mm, pore size 0.45 µm, e. g. Acrodisc for ion chromatography, from Pall GmbH, D-63303 Dreieich

- Glass autosampler vials, e.g. from Merck, D-64271 Darmstadt
- Screw caps with a hole and silicone/PTFE septa, e.g. from Merck
- Ultrapure water system for the preparation of ultrapure water, e.g. NANOpure ultrapure water system from Barnstead, supplier in Germany: Wilhelm Werner GmbH, D-51381 Leverkusen
- Ultrasonic bath
- HPIC apparatus with conductivity detector, e.g. from Merck Hitachi, D-64271 Darmstadt

The volumetric flasks and polyethylene bottles listed are cleaned in a dishwasher (before first use also), rinsed with ultrapure water and subsequently dried in a drying cabinet.

1.2 Chemicals and Solutions

Sulfate standard, w = 1.000 $\pm$ 0.002 g/L, e.g. from Merck
Sodium carbonate, anhydrous, analytical grade, e.g. from Fluka, Sigma-Aldrich Chemie GmbH, D-82024 Taufkirchen
Sodium hydrogen carbonate, analytical grade, e.g. from Fluka, Sigma-Aldrich Chemie GmbH
Ultrapure water

Eluent stock solution:	Solution of 0.62 mol/L sodium carbonate and 0.069 mol/L sodium hydrogen carbonate in ultrapure water. 13.14 g sodium carbonate and 1.15 g sodium hydrogen carbonate are dissolved in ultrapure water in a 200 mL volumetric flask. The flask is then filled to the mark with ultrapure water and shaken.
Eluent:	Solution of 3.1 mmol/L sodium carbonate and 0.35 mmol/L sodium hydrogen carbonate in ultrapure water. 10 mL eluent stock solution is added with a pipette into a 2-litre volumetric flask containing a few millilitres of ultrapure water. The flask is then filled to the mark with ultrapure water and shaken.
Stock solution:	Solution of 200 mg/L sulfate in ultrapure water. 4.0 mL sulfate standard solution is pipetted into a 20 mL volumetric flask. The flask is then filled to the mark with eluent and shaken. The solution is stable at room temperature for at least 4 weeks.
Calibration solutions:	Solutions of 0.8 to 20 mg/L sulfate in eluent. The volumes of calibration stock solution specified in Table 1 are pipetted into separate 20 mL volumetric flasks, each containing a few millilitres of eluent. The flasks are then filled to the mark with eluent and shaken.

For a 420-litre air sample volume, these solutions cover a sulfuric acid concentration range from approx. 0.01 to 0.19 mg/m^3.
The calibration solutions are stable at room temperature for at least 4 weeks.
Calibration is checked on working days e.g. with a solution containing sulfate at a concentration of 9 mg/L.

Table 1.

Calibration solution No.	Volume of the stock solution μL	Concentration mg SO$_4^{2-}$/L	Concentration mg/m^3 H$_2$SO$_4$
1	80	0.8	0.01
2	200	2.0	0.02
3	400	4.0	0.04
4	600	6.0	0.06
5	900	9.0	0.09
6	1200	12.0	0.12
7	1400	14.0	0.14
8	1600	16.0	0.16
9	1800	18.0	0.18
10	2000	20.0	0.19

2 Sampling

The filter holder is fitted with a binder-free quartz fibre filter and connected to the pump. To take into account the definition of inhalable dust fraction [2] air is drawn through the filter holder with a flow-regulated pump at a flow rate of 210 L/hour over a period of 2 hours.
The filters should be transferred to a polyethylene bottle and covered with 4 mL eluent immediately after sampling. The polyethylene bottle should be sealed with a screw cap.

3 Analytical determination

3.1 Sample preparation and analysis

After stabilisation in elution solution, the sample is treated in a ultrasonic bath for 15 minutes and allowed to cool to room temperature for about 30 minutes. As much of the elution solution as possible (3.2–3.5 mL) is drawn into a disposable 5-mL syringe. The solution is filtered through a disposable filter into an autosampler vial as well as into another vial in case repeat measurements or sample dilution becomes necessary. 50 μL of this solution is injected into the ion chromatograph.
Sample preparation must be carried out as described in order to attain the recoveries given in Section 5.1.

3.2 Operating conditions for ion chromatography

The method was characterised under the following experimental conditions:

Apparatus:	Ion chromatography system, consisting of the LaChrom L-7100 low-pressure gradient pump, L-7200 autosampler, L-7350 column oven and L-7470 conductivity detector and the D-7000 HSM chromatography software from Merck Hitachi, D-64271 Darmstadt
Precolumn:	IonPac AG12A (4 × 50 mm), from Dionex GmbH, D-65510 Idstein
Column:	IonPac AS12A (4 × 200 mm), from Dionex GmbH
Suppressor:	ASRS Ultra Anion Self-Regenerating Suppressor, 4 mm, from Dionex GmbH
Eluent:	Na_2CO_3 (3.1 mmol/L)/$NaHCO_3$ (0.35 mmol/L) solution
Flow rate:	1.5 mL/min
Injection volume:	50 µL
Column temperature:	35 °C

4 Calculations

4.1 Calibration

Aliquots of 50 µL of each of the calibration solutions described in Section 1.2 are injected into the ion chromatograph. The peak areas determined are plotted against the weights of sulfate contained in the corresponding solutions in order to construct the calibration curve. The calibration curve is linear in the specified concentration range.

4.2 Calculation of the analytical result

The concentration of sulfuric acid in the air sample is calculated in mg/m^3 according to Equation (1):

$$c_w = 1.021 \frac{w}{V \cdot \eta} \tag{1}$$

where:

c_w	is the concentration of sulfuric acid in the air sample in mg/m^3
w	is the weight of sulfate in µg per sample as determined from the calibration curve
1.021	is the conversion factor for sulfate to sulfuric acid
V	is the volume of the air sample in litres
η	is the recovery

5 Reliability of the method

The method was developed in accordance with DIN EN 32645 and DIN EN 482 [3, 4].

5.1 Accuracy and recovery

Accuracy in the minimum range of measurement according to DIN EN 482 and recoveries were determined for four different concentrations (see Table 2).
Recovery was determined by spiking 21 µL, 42 µL, 168 µL and 420 µL of the stock solution (see Section 1.2) onto individual quartz fibre filters. The filters were subsequently dried and 420 litres of laboratory air were drawn through them (30–40% relative air humidity). The filters were then immediately transferred to separate polyethylene bottles and covered with 4 mL eluent. The polyethylene bottles were then sealed with screw caps. The remaining steps of sample preparation were carried out as described in Section 3.1. Each concentration was determined in replicates of ten.
The concentration range covered in this way was from 0.01 to 0.2 mg/m^3.
The accuracy and recovery results are presented in Table 2.

Table 2.

Weight µg	Concentration* mg/m^3	Relative standard deviation %	Recovery
4.2	0.01	2.6	1.03
8.4	0.02	1.6	1.00
33.6	0.08	0.5	1.00
84.0	0.20	0.6	0.97

* for an air sample volume of 420 litres

5.2 Limit of quantification

The absolute limit of quantification for sulfuric acid under the stated conditions of analysis is 50 ng. For a 420-litre air sample, 4 mL elution solution and 50 µL injection volume the relative limit of quantification is 0.01 mg/m^3.
It was determined using the linear calibration method according to DIN EN 32645 [3].

5.3 Selectivity

The selectivity of the chromatographic method depends above all on the type of column and the separation conditions used. The separation conditions described here have proved reliable in practice.

5.4 Shelf-life

In order to check the shelf-life, 12 filters were spiked with different weights of sulfuric acid, corresponding to high (0.2 mg/m^3) and low (0.02 mg/m^3) concentrations of sulfuric acid in the air. For sample treatment see Sections 3.1 and 5.1. The elution solutions were stored at room temperature for 7 days and subsequently kept in a refrigerator at 4 °C. The solutions were then analysed in duplicate on days 1, 5, 8, 15, 22 and 29. No decrease in concentration was observed.

5.5 Discussion

The working range of the method described can be extended to concentrations of up to 2 mg/m^3.

The method is suitable for working areas in which sulfuric acid aerosols occur.

In working areas where solutions are used which contain sulfates in addition to sulfuric acid, the composition of the solution needs to be taken into account when the result of the measurement is calculated [5].

In working areas where oleum (sulfuric acid enriched with SO$_3$) is handled, determination of the total concentration of SO$_3$ and sulfuric acid requires a different method (see Sulfuric acid, Method No. 2 or BIA Arbeitsmappe Blatt 8580 Verfahren Nr. 3 [6]).

Sulfur dioxide is not detected by the sampling procedure described.

The present method also permits the determination of phosphoric and oxalic acid concentrations in the air in working areas.

6 References

[1] *BIA-Arbeitsmappe,* Messung von Gefahrstoffen, Blatt Nr. 3010. Erich Schmidt Verlag, Bielefeld, 27. Lieferung 2001

[2] *Deutsches Institut für Normung e.V.* (DIN) (1993) DIN EN 481 – Arbeistplatzatmosphäre – Festlegung der Teilchengrößenverteilung zur Messung luftgetragener Partikel. Beuth Verlag, Berlin

[3] *Deutsches Institut für Normung e.V.* (DIN) (1994) DIN 32645 – Chemische Analytik – Nachweis-, Erfassungs- und Bestimmungsgrenze. Beuth Verlag, Berlin

[4] *Deutsches Institut für Normung e.V.* (DIN) (1994) DIN EN 482 – Arbeitsplatzatmosphäre – Allgemeine Anforderungen an Verfahren für Messung von chemischen Arbeitsstoffen. Beuth Verlag, Berlin

[5] *Bundesministerium für Arbeit und Sozialordnung* (2003) TRGS 901, Lfd-Nr. 104, BArbBl 2/2003: 92–96

[6] *BIA Arbeitsmappe,* Messung von Gefahrstoffen, Blatt 8580, Verfahren Nr. 3. Erich Schmidt Verlag, Bielefeld, 27. Lieferung 2001

Author: *D. Breuer*

Federation of the Employment Accidents Insurance Institutions of Germany
(Hauptverband der Berufsgenossenschaften)
Centre for Accident Prevention and Occupational Medicine
Alte Heerstraße 111, 53757 Sankt Augustin
Expert Committee Chemistry

Carcinogenic substances	Order number:	BGI 505-71E
Established methods	Issue:	August 2001

Method for the determination of sulfuric acid or oleum

Method tested and recommended by the Berufsgenossenschaften for the determination of sulfuric acid or oleum in working areas after discontinuous sampling.
Both personal and stationary sampling can be conducted for the assessment of working areas.

Sampling with a pump and absorption in a sodium carbonate/sodium hydrogen carbonate solution, ion chromatography (HPIC) after elution.
Sulfuric acid-2-Ion chromatography
Sampling with a washing bottle
(Issue: August 2001)

Name: Sulfuric acid

CAS No.: 7664-93-9

Molecular formula: H_2SO_4

Molar mass: 98.07 g/mol

The MAK-Collection Part III: Air Monitoring Methods, Vol. 9. DFG, Deutsche Forschungsgemeinschaft
Copyright © 2005 WILEY-VCH Verlag GmbH & Co. KGaA, Weinheim
ISBN: 3-527-31138-6

Sampling with a pump and absorption in a sodium carbonate/sodium hydrogen carbonate solution, ion chromatography after elution

Summary

This method permits the determination of sulfuric acid concentrations in working areas averaged over the sampling time after personal or stationary sampling. Sulfur trioxide is also detected under the sampling conditions described.

Principle:	A pump is used to draw a measured volume of air through a washing bottle filled with sodium carbonate/sodium hydrogen carbonate solution. The absorbed sulfuric acid is determined by ion chromatography.
Technical data:	
Limit of quantification:	absolute: 5 ng sulfuric acid
	relative: 0.002 mg/m^3 sulfuric acid for a 560-litre air sample volume, 10 mL absorbent volume and 50 µL injection volume
Selectivity:	The selectivity of the chromatographic method depends above all on the type of column and the separation conditions used. The separation conditions described here have proved reliable in practice.
Advantages:	Personal sampling and selective determinations possible
Disadvantages:	No indication of peak concentrations
Apparatus:	Pump
	Gas meter or volumetric flow meter
	Washing bottle
	Ion chromatograph with conductivity detector

Detailed description of the method

Contents

1 Equipment, chemicals and solutions

1.1 Equipment

For sampling:
- Pump for personal air sampling, suitable for a flow rate of 70 L/hour, e. g. Gilian PP5-Ex, supplier in Germany: e. g. DEHA Haan + Wittmer GmbH, D-71292 Friolzheim
- Leak-proof washing bottle „absorber B 70 according to BIA", available from e. g. GSM GmbH, D-41469 Neuss
- Gas meter or volumetric flow meter

For sample preparation and analysis:
- Volumetric flasks, 10 to 1000 mL
- Adjustable piston pipettes, 10 µL to 5000 µL, e. g. from Eppendorf AG, D-22339 Hamburg
- Autosampler vials, e. g. Polyvial 1 mL from Dionex GmbH, D-65510 Idstein
- Ultrapure water system for the preparation of ultrapure water, e. g. NANOpure ultrapure water system from Barnstead, supplier in Germany: Wilhelm Werner GmbH, D-51381 Leverkusen
- Ultrasonic bath
- HPIC apparatus with conductivity detector, e. g. from Dionex GmbH

1.2 Chemicals and Solutions

Sodium sulfate, purity $\geq$ 99%, e.g. from Merck, D-64271 Darmstadt
Sodium carbonate, anhydrous, >99%, e.g. from Fluka, Sigma-Aldrich Chemie GmbH, D-82024 Taufkirchen
Sodium hydrogen carbonate, purity $\geq$ 99.5%, e.g. from Fluka, Sigma-Aldrich Chemie GmbH
Sodium oxalate, 99.9%, e.g. from Merck
Ultrapure water

Eluent stock solution A:	Solution of 0.5 mol/L sodium carbonate in ultrapure water. 26.49 g Na_2CO_3 is weighed into a 500 mL volumetric flask, which is then filled to the mark with ultrapure water and shaken.
Eluent stock solution B:	Solution of 0.5 mol/L sodium hydrogen carbonate in ultrapure water. 21.0 g $NaHCO_3$ is weighed into a 500 mL volumetric flask, which is then filled to the mark with ultrapure water and shaken.
Eluent:	Solution of 3.5 mmol/L sodium carbonate and 0.5 mmol/L sodium hydrogen carbonate in ultrapure water. 7 ml eluent stock solution A and 1 mL eluent stock solution B are pipetted into a 1000 mL volumetric flask containing a few millilitres of ultrapure water. The flask is then filled to the mark with ultrapure water and shaken.
Internal standard stock solution:	Solution of 1 g/L oxalate in eluent. 152 mg sodium oxalate is weighed into a 100 mL volumetric flask to the nearest 0.1 mg. The flask is then filled to the mark with eluent and shaken.
Absorption solution:	Solution of 3.5 mmol/L sodium carbonate, 0.5 mmol/L sodium hydrogen carbonate and 0.01 g/L oxalate in ultrapure water. 5 mL internal standard stock solution is pipetted into a 500 mL volumetric flask containing a few millilitres of eluent. The flask is then filled to the mark with eluent and shaken.
Calibration stock solution:	Solution of 2 g/L sulfate and 0.01 g/L oxalate in eluent. 295.8 mg sodium sulfate is weighed into a 100 mL volumetric flask to the nearest 0.1 mg. The flask is then filled to the mark with absorption solution and shaken.
Calibration solutions:	Solutions of 0.2 to 12.5 mg/L sulfate and 0.01 g/L oxalate in eluent. The volumes of calibration stock solution specified in Table 1 are pipetted into separate 20 mL volumetric flasks, each containing a few millilitres of absorption solution. The flasks are then filled to the mark with absorption solution and shaken. For a 560-litre air sampling volume, these solutions cover a sulfuric acid concentration range of approx. 0.004 to 0.23 mg/m^3 air.

Table 1.

Calibration solution No.	Volume of the stock solution μL	Concentration mg SO_4^{2-}/L	Concentration mg/m^3 H_2SO_4
1	2	0.2	0.004
2	25	2.5	0.05
3	50	5.0	0.09
4	75	7.5	0.14
5	100	10.0	0.18
6	125	12.5	0.23

Validation stock solution: Solution of 0.12 g/L sulfate and 0.01 g/L oxalate in eluent. 17.1 mg sodium sulfate is weighed into a 100 mL volumetric flask to the nearest 0.1 mg. The flask is then filled to the mark with absorption solution and shaken.

The calibration solutions are stable at room temperature for at least 4 weeks. Calibration is checked on working days e. g. with a solution containing sulfate at a concentration of 5 mg/L.

2 Sampling

The absorber B 70 is filled with 10 mL absorption solution and connected to the pump. The pump and absorber are carried by a person during working hours or used in a stationary position. The flow rate is set at 1.16 L/min. Under these conditions, sampling is in accordance with the definition of inhalable dust fraction [1]. A sampling time of eight hours then corresponds to an air sample volume of about 560 litres.

3 Analytical determination

3.1 Sample preparation and analysis

The contents of the absorber B70 are quantitatively transferred to a 10 mL volumetric flask, which is then filled to the mark with eluent (sample solution). After brief shaking, an aliquot is transferred with a pipette to a sample vial. 50 µL of this solution is injected into the ion chromatograph.

3.2 Operating conditions for ion chromatography

The method was characterised under the following experimental conditions:

Apparatus:	DIONEX DX 120 Ion Chromatograph with autosampler
Detector:	DIONEX CDM-1 Conductivity Detector
Precolumn:	IonPac AG14 (4 × 50 mm), from Dionex GmbH, D-65510 Idstein
Column:	IonPac AS14 (4 × 200 mm), from Dionex GmbH
Suppressor:	ASRS 1, from Dionex GmbH
Eluent:	Na_2CO_3 (c = 3.5 mmol/L)/$NaHCO_3$ (c = 0.5 mmol/L) solution
Flow rate:	1.2 mL/min
Injection volume:	50 µL
Column temperature:	Room temperature

4 Calculations

4.1 Calibration

Aliquots of 50 µL of each of the calibration solutions described in Section 1.2 are injected into the ion chromatograph. In order to construct the calibration curve, the peak area ratios determined for the sulfate ions and oxalate ions as internal-standard are plotted against the weight ratios of sulfate and oxalate ions contained in the individual solutions.

4.2 Calculation of the analytical result

The concentration by weight of sulfuric acid in the air sample is calculated in mg/m^3 according to Equation (1):

$$c_w = 1.021 \frac{w}{V \cdot \eta} \tag{1}$$

where:

c_w	is the concentration by weight of sulfuric acid in the air sample in mg/m^3
w	is the weight of sulfate in µg per sample as determined from the calibration curve
1.021	is the conversion factor for sulfate to sulfuric acid
V	is the volume of the air sample in litres
η	is the recovery

5 Reliability of the method

5.1 Accuracy and recovery

Accuracy in the minimum range of measurement according to DIN EN 482 [2] and recoveries were determined for three different concentrations (see Table 2).
Recovery was determined by pipetting 49, 490 and 980 µL of the validation stock solution described in Section 1.2 into 10 mL volumetric flasks, which were then filled to
the mark with absorption solution and shaken. The solutions were each transferred to
an absorber B 70. Subsequently, 560 litres of laboratory air was drawn through the absorber at a flow rate of 1.16 L/min and sample preparation carried out as described in
Section 3.1. Each concentration was determined in replicates of six.
The concentration range covered in this way was from 0.01 to 0.21 mg/m^3.
The accuracy and recovery results are presented in Table 2.

Table 2.

Concentration* mg/m^3	Relative standard deviation %	Recovery
0.01	4.3	1.06
0.10	3.2	1.09
0.21	3.2	1.07

* for an air sample volume of 560 litres

5.2 Limit of quantification

The limit of quantification was determined from the signal/noise ratio (10:1) of the
chromatogram baseline.
The absolute limit of quantification for sulfate is 0.005 µg. The relative limit of quantification, expressed in terms of sulfuric acid, is 0.002 mg/m^3 for 560-litre air sample, 10
mL sample solution and 50 µL injection volume.

5.3 Selectivity

Any sulfur dioxide present in working areas is also collected during sampling, oxidised
and analysed as sulfate.
The selectivity of the chromatographic method depends above all on the type of column
and the separation conditions used. The separation conditions described here have
proved reliable in practice.

5.4 Shelf-life

The shelf-life of the samples is at least 4 weeks in the refrigerator at 4 °C.

5.5 Discussion

The method is suitable for working areas in which sulfuric acid aerosols occur.
In working areas where solutions are used which contain sulfates in addition to sulfuric acid, the composition of the solution needs to be taken into account when the result of the measurement is calculated [3].
In working areas where oleum (sulfuric acid enriched with SO_3) is handled, the total concentration of SO_3 and sulfuric acid is determined.
The present method also permits the determination of the concentrations of phosphoric acid, hydrochloric acid, nitric acid and hydrobromic acid in the air in working areas.

6 References

[1] *Deutsches Institut für Normung e.V.* (DIN) (1993) DIN EN 481 – Arbeitsplatzatmosphäre – Festlegung der Teilchengrößenverteilung zur Messung luftgetragener Partikel. Beuth-Verlag, Berlin.
[2] *Deutsches Institut für Normung e.V.* (DIN) (1994): DIN EN 482 – Arbeitsplatzatmosphäre – Allgemeine Anforderungen an Verfahren für Messung von chemischen Arbeitsstoffen. Beuth-Verlag, Berlin.
[3] *Bundesministerium für Arbeit und Sozialordnung* (2003) TRGS 901, Lfd-Nr. 104, BArbBl 2/2003: 92–96.

Author: *W. Krämer*

Federation of the Employment Accidents Insurance Institutions of Germany
(Hauptverband der Berufsgenossenschaften)
Centre for Accident Prevention and Occupational Medicine
Alte Heerstraße 111, 53757 Sankt Augustin
Expert Committee Chemistry

Carcinogenic substances	Order number: BGI 505-69E
Established methods	Issue: April 2001

Method for the determination of thiourea

Method tested and recommended by the Berufsgenossenschaften for the determination of thiourea in working areas after discontinuous sampling.
Both personal and stationary sampling can be conducted for the assessment of working areas.

Sampling with a pump and collection on a filter, high-performance liquid chromatography (HPLC) after elution.
Thiourea-1-HPLC
(Issue: April 2001)

Name:	Thiourea
CAS No.:	62-56-6
Molecular formula:	CH_4N_2S
Molar mass:	76.12 g/mol

The MAK-Collection Part III: Air Monitoring Methods, Vol. 9. DFG, Deutsche Forschungsgemeinschaft
Copyright © 2005 WILEY-VCH Verlag GmbH & Co. KGaA, Weinheim
ISBN: 3-527-31138-6

Sampling with a pump and collection on a filter, HPLC after elution

Summary

This method permits the determination of thiourea concentrations in working areas averaged over the sampling time after personal or stationary sampling.

Principle: A pump is used to draw a measured volume of air through a glass fibre filter, taking into account the definition of inhalable dust fraction [1]. The collected thiourea is eluted with water and determined by liquid chromatography.

Technical data:

Limit of quantification: absolute: 0.58 ng

relative: personal sampling: 1.73 $\mu g/m^3$ for a 1.68 m^3 air sample, 100 mL elution solution and an injection volume of 20 μL

stationary sampling: 0.64 $\mu g/m^3$ for a 45 m^3 air sample, 1000 mL elution solution and 20 μL injection volume

Selectivity: The selectivity of the method depends above all on the type of column and chromatography conditions used. The separation conditions described here have proved reliable in practice.

Advantages: Personal sampling and selective determinations possible

Disadvantages: No indication of peak concentrations

Apparatus: Pump
Gas meter or volumetric flow meter
Binder-free glass fibre filter and filter holder
Liquid chromatograph with UV/VIS detector

Detailed description of the method

Contents

1 Equipment, chemicals and solutions

1.1 Equipment

For sampling:

Personal sampling:
- Pump, suitable for flow rates of 3.5 L/min, e.g. Gilian PP5-Ex, suppliers in Germany: e.g. DEHA Haan + Wittmer GmbH, D-71292 Friolzheim
- Gas meter or volumetric flow meter
- Filter holder GSP, e.g. from DEHA Haan + Wittmer GmbH
- Glass fibre filter, binder-free, diameter 37 mm, e.g. MN 85/90 BF, from Macherey-Nagel GmbH & Co. KG, D-52355 Düren
- Ultrapure water system, e.g. NANOpure ultrapure water system from Barnstead, supplier in Germany: Wilhelm Werner GmbH, D-51381 Leverkusen

Stationary sampling:
- Sampling system (pump, filter holder, gas meter), suitable for flow rates of 22.5 m^3/hour, e.g. Gravicon VC 25 G, e.g. from Juwe Laborgeräte GmbH, D-41751 Viersen
- Glass fibre filter, binder-free, diameter 150 mm, e.g. 13400B-150K, from Sartorius AG, D-37075 Göttingen

For sample preparation and analysis:
– Volumetric flasks, 1000 mL, 500 mL, 250 mL, 100 mL
– HPLC apparatus
– UV/VIS detector
– Data system
– Ultrasonic bath
– Pipettes, 50 mL, 20 mL, 10 mL
– Measuring pipettes, suitable for volumes between 50 μL and 1000 μL
Injection loop, 20 μL

1.2 Chemicals and Solutions

Thiourea:	≥ 99%, e.g. from Degussa AG, D-83303 Trostberg.
Water:	HPLC grade
Standard solution:	Solution of thiourea in water, 10 mg/L. Approximately 10 mg thiourea is weighed into a 1000-mL volumetric flask to the nearest 0.1 mg, and dissolved in water. The flask is then filled to the mark with water and shaken.
Calibration solutions:	Aqueous solutions containing thiourea at 5, 2, 1, 0.4, 0.2 and 0.1 mg/L. Portions of 50, 20 and 10 mL of stock solution are pipetted into three 100 mL volumetric flasks and 10 mL portions are pipetted into a 250 mL, a 500 mL and a 1000 mL volumetric flask. The flasks are then filled to the mark with water and shaken.

2 Sampling

2.1 Personal sampling

The filter holder is fitted with a binder-free glass fibre filter and connected to the pump. The pump and filter holder are carried by a person during working hours. To take into account the definition of inhalable dust fraction [1] the flow rate is set at 3.5 L/min. A sampling time of eight hours then corresponds to a sample volume of 1.68 m^3.

2.2 Stationary sampling

The sampling system is fitted with a binder-free glass fibre filter and placed in a stationary position. To take into account the definition of inhalable dust fraction [1] the flow rate is set at 22.5 m^3/hour. A sampling time of two hours then corresponds to a sample volume of 45 m^3.

3 Analytical determination

3.1 Sample preparation and analysis

For processing after personal sampling, the filter is placed in a 100 mL volumetric flask and 80 mL water is added. For stationary sampling, a 1000 mL volumetric flask is used and 800 mL water is added. The volumetric flasks are subsequently placed in the ultrasonic bath for 10 minutes, allowed to cool to room temperature and filled to the mark with water (sample solution). Aliquots of 20 µL of these solutions are injected into the liquid chromatograph. Thiourea is detected at a wavelength of 245 nm.

3.2 Operating conditions for HPLC

The method was characterised under the following experimental conditions:

Apparatus: Liquid chromatograph HP1100, from Hewlett Packard,
                      D-76337 Waldbronn
Column: Nucleosil 5 C-18 (length 250 mm, internal diameter: 4.6 mm),
                      Macherey-Nagel GmbH & Co. KG, D-52355 Düren
Eluent: 100% water
Flow rate: 0.8 mL/min
Injection volume: 20 µL
Detection wavelength: 245 nm
Column temperature: Room temperature

4 Calculations

4.1 Calibration

Aliquots of 20 µL of each of the calibration solution are injected into the chromatograph and chromatograms are recorded. The peak areas determined are plotted against the weights of thiourea contained in the corresponding calibration solutions in order to construct the calibration curve. It is linear in the specified concentration range from 0.1 to 5.0 mg/L.

4.2 Calculation of the analytical result

The concentration by weight of thiourea in the air sample is calculated in mg/m^3 according to equation (1):

$$c_w = \frac{w}{V \cdot \eta}$$

(1)

where:

c_w is the concentration by weight of thiourea in the air sample in mg/m³
w is the weight of thiourea in the sample solution in mg as determined from the cali-
      bration curve
V is the volume of the air sample in m³
η is the recovery.

5 Reliability of the method

5.1 Accuracy and recovery

Personal sampling:
To determine recovery, aliquots of 50 μL and 500 μL of a solution containing 3.2 g
thiourea per litre of water were each spiked onto a binder-free filter for personal sampling.
Subsequently, 1.68 m³ of air was drawn through each filter as described in Section 2.1.
With this air sample volume, the spiked amounts correspond to 0.1 and 1 mg/m³.
Each concentration was determined in replicates of six. The respective relative standard
deviations were 3.9 and 1.8%. The respective recoveries for 0.1 and 1 mg/m³ were 0.78
and 0.96.

Stationary sampling:
To determine recovery, aliquots of 0.1 mL and 1 mL of a solution containing 46.8 g
thiourea per litre of water were each spiked onto a binder-free filter for stationary sampling. Subsequently, 45 m³ of air was drawn through each filter as described in Section 2.2. With this air sample volume, the spiked amounts correspond to 0.1 and 1 mg/m³.
Each concentration was determined in replicates of six. The respective relative standard
deviations were 1.6 and 1.3%. The respective recoveries for 0.1 and 1 mg/m³ were 0.94
and 0.87.

5.2 Limit of quantification

The limit of quantification is 0.029 mg/L thiourea. This is calculated from the standard
deviation of six replicate determinations of a 0.1 mg/L standard solution according to
equation (2):

$$LOQ = 10 \cdot s$$

(2)

where:
LOQ   is the limit of quantification
s is the standard deviation of six replicates

The absolute limit of quantification is 0.58 ng for an injection volume of 20 μL. This corresponds to 2.9 μg/sample for 100 mL of sample solution (personal sampling), or 29 μg/sample for 1000 mL of sample solution (stationary sampling).

For personal sampling, the relative limit of quantification is 1.73 μg/m^3 for an air sample volume of 1.68 m^3, 100 mL sample solution and an injection volume of 20 μL. For stationary sampling, the relative limit of quantification is 0.64 μg/m^3 for an air sample volume of 45 m^3, 1000 mL sample solution and an injection volume of 20 μL.

5.3 Selectivity

The selectivity of the method depends above all on the type of column and chromatography conditions used. The separation conditions described here have proved reliable in practice.

5.4 Shelf-life

In order to check the shelf-life, six filters were spiked with 1 mg/m^3 thiourea (see Section 5.1) and stored at room temperature. Three filters were eluted and analysed for thiourea by liquid chromatography after 14 days, the remaining three after 28 days. No decrease in concentration was observed.

6 References

[1] *Deutsches Institut für Normung e.V.* (DIN) (1993): DIN EN 481 – Arbeitsplatzatmosphäre – Festlegung der Teilchengrößenverteilung zur Messung luftgetragener Partikel. Beuth-Verlag, Berlin

Author: *U. Rust*

Federation of the Employment Accidents Insurance Institutions of Germany
(Hauptverband der Berufsgenossenschaften)
Centre for Accident Prevention and Occupational Medicine
Alte Heerstraße 111, 53757 Sankt Augustin
Expert Committee Chemistry

Carcinogenic substances	Order number: BGI 505-65E
Established methods	Issue: March 2000

Method for the determination of trichloroethene and tetrachloroethene

Method tested and recommended by the Berufsgenossenschaften for the determination of trichloroethene (TCE) and tetrachloroethene (PCE) in working areas.
Both personal and stationary sampling can be conducted for the assessment of working areas.

Sampling with a pump and adsorption on activated carbon
Gas chromatography after desorption
Trichloroethene/Tetrachloroethene-1-GC
(Issue: March 2000)

	TCE	**PCE**
IUPAC name:	Trichloroethene (Trichloroethylene)	Tetrachloroethene (Tetrachloroethylene) (Perchloroethylene)
CAS No.:	79-01-6	127-18-4
Molecular formula:	C_2HCl_3	C_2Cl_4
Molar mass:	131.39 g/mol	165.82 g/mol

The MAK-Collection Part III: Air Monitoring Methods, Vol. 9. DFG, Deutsche Forschungsgemeinschaft
Copyright © 2005 WILEY-VCH Verlag GmbH & Co. KGaA, Weinheim
ISBN: 3-527-31138-6

Summary

This method permits the determination of TCE and PCE concentrations in working areas averaged over the sampling time after personal or stationary sampling.

Principle:	A pump is used to draw a measured air volume through a tube filled with activated carbon. The adsorbed TCE and/or PCE is desorbed with carbon disulfide and determined by gas chromatography.

Technical data:

Limit of quantification:	absolute: 8 ng TCE or PCE
	relative: 4.2 mg/m^3 0.76 mL/m^3 (ppm) for TCE and 4.2 mg/m^3 0.60 mL/m^3 (ppm) for PCE, both determined for a 9.6-litre air sample, 5 mL desorption solution and 1 µL injection volume
Selectivity:	In the presence of interfering components, the values determined may be too high. Interference can generally be eliminated by choosing a column with different separation characteristics.
Advantages:	Personal sampling and selective determinations possible
Disadvantages:	No indication of peak concentrations
Apparatus:	Pump
	Gas meter or volumetric flow meter
	Tubes filled activated carbon
	Gas chromatograph with flame ionisation detector (FID)

Detailed description of the method

Contents

1 Equipment, chemicals and solutions

1.1 Equipment

For sampling:
– Pump, suitable for flow rates of 80 mL/min, e.g. PP1 from Gilian, suppliers in Germany: DEHA Haan + Wittmer GmbH, D-71292 Friolzheim
– Gas meter or volumetric flow meter
– Adsorption tubes filled with activated carbon (standardised, consisting of two sections filled with about 300 mg (sampling section) and 800 mg (backup section) of activated carbon and separated by a porous polymer material), type B, e.g. from Dräger, D-23560 Lübeck
– Caps for the opened activated carbon tubes

For sample preparation and analysis:
– Volumetric flasks, 10 mL, 25 mL, 50 mL
– Sample vials with PTFE[1]-coated septa and caps, approx. 10 mL
– Pipettes, 2 mL, 2.5 mL, 5 mL, 10 mL
– Mechanical shaker
– Syringes, 10 μL, 50 μL, 100 μL
– Gas chromatograph with FID
– Data system

---

1 Polytetrafluoroethylene.

1.2 Chemicals and Solutions

Trichloroethene, purity $\geq 99\%$
Tetrachloroethene, purity $\geq 99\%$
Carbon disulfide, purity $\geq 99.9\%$, e.g. Uvasol, low-benzene, from Merck, D-64271
Darmstadt, Catalogue No. 102213.

Stock solution:	Solution of approx. 64 mg/mL TCE and approx. 130 mg/mL PCE in carbon disulfide. Approximately 6600 mg (approx. 4 mL) tetrachloroethene and then approx. 3200 mg trichloroethene (approx. 2.2 mL) are precisely weighed into a 50 mL volumetric flask. The flask is then filled to the mark with carbon disulfide and shaken.
Calibration solutions:	Solutions of approx. 0.032, 0.16, 0.32 and 0.64 mg/mL TCE and approx. 0.065, 0.325, 0.65 and 1.3 mg/mL PCE in carbon disulfide. Syringes are used to inject 5, 25, 50 and 100 µL of stock solution into four 10 mL volumetric flasks, each containing approx. 5 mL carbon disulfide. The flask is then filled to the mark. For a 9.6-litre air sample volume and 5 mL sample solution, these solutions cover respective TCE and PCE concentration ranges of approx. 17 to 330 mg/m^3 and approx. 34 to 690 mg/m^3 air.
Gases for gas chromatography:	Helium, purity $\geq 99.999\%$ Hydrogen, purity $\geq 99.995\%$ Synthetic air, hydrocarbon-free Nitrogen, purity $\geq 99.999\%$

2 Sampling

An activated carbon tube is opened and connected to the pump such that the sampling section will be loaded first. The flow rate is set at 80 mL/min. A sampling time of 2 hours then corresponds to an air sample volume of 9.6 litres. The pump and tube are carried by a person during working hours or used in a stationary position. On completion of sampling, the tube is tightly sealed. The method was tested up to an air sample volume of 10 litres at a maximum flow rate of 80 mL/min.

3 Analytical determination

3.1 Sample preparation and analysis

The contents of the loaded activated charcoal tube are transferred – sampling and backup sections separately – to 10 mL sample vials and 5 mL carbon disulfide is added to each vial. The vials are then immediately sealed and shaken for 30 minutes.
Aliquots of 1 µL of the desorption solutions are injected into the gas chromatograph.
When an autosampler is used, samples are first transferred to autosampler vials. The quantitative analysis of the chromatograms is performed by the external standard method.

3.2 Operating conditions for gas chromatography

The method was characterised under the following experimental conditions:

Apparatus:	Hewlett-Packard Model 6890 gas chromatograph with FID, autosampler (6890 Series Injector) and CIS 3 cold injection system from Gerstel, D-45473 Mülheim.
Column:	Quartz capillary, stationary phase: Ultra 2 (5% diphenylpolysiloxane, 95% dimethylpolysiloxane), length: 25 m, internal diameter: 0.20 mm, film thickness: 0.33 µm.
Temperatures:	Cold injection system:
	From 20 °C to 180 °C at a rate of 12 °C/s, 0.5 min at 180 °C
	Furnace:
	Initial temperature: 35 °C, 3 minutes isothermal
	Heating rate 1: 5 °C/min up to 70 °C
	Heating rate 2: 40 °C up to 140 °C
	140 °C, 1 minute isothermal
	Detector:
	250 °C
Type of injection:	Cold injection
Carrier gas:	Helium, 1.0 mL/min
Split ratio:	20 : 1
Detector gases:	Hydrogen, 40 mL/min
	Synthetic air, 450 mL/min
	Nitrogen (make-up gas), 45 mL/min
Injection volume:	1 µL

4 Calculations

4.1 Calibration

Aliquots of 1 μL of each of the calibration solutions described in Section 1.2 are injected into the gas chromatograph. The peak areas determined are plotted against the corresponding TCE and PCE concentrations contained in the calibration solutions in order to construct the calibration curves. They are linear under the conditions described.

4.2 Calculation of the analytical result

The peak areas for TCE or PCE are determined, and the corresponding weight of analyte in the sample is read from the calibration curve in μg. If the weight of analyte adsorbed in the backup section exceeds by more than 30% that adsorbed in the collection section, sampling must be repeated with a smaller air sample volume; if not possible, the two values are added.

The concentration by weight in the air sample in mg/m³ is calculated according to Equation (1):

$$c_w = \frac{W}{V \cdot \eta}$$
(1)

The concentration by volume c_v in mL/m³ at 20 °C and 1013 hPa is calculated from c_w as follows (Equation 2 and 3):

for TCE: $c_v = 0.183 \cdot c_w$ (2)
for PCE: $c_v = 0.145 \cdot c_w$ (3)

where:

c_w is the TCE or PCE concentration by weight in the air sample, given in mg/m³
c_v is the TCE or PCE concentration by volume in the air sample, given in mL/m³ (ppm)
w is the weight of TCE or PCE in the desorption solution(s) in μg as determined from the appropriate calibration curve
V is the air sample volume in litres
η is the recovery

5 Reliability of the method

5.1 Accuracy and recovery

The following spiking solutions were prepared in order to determine the relative standard deviation of the method:

Spiking solution 1:   7718 mg TCE (approx. 5 mL) and 16380 mg PCE (approx.10 mL) were weighed into a 25 mL volumetric flask, and methanol was added to the mark.

Spiking solution 2:   10 mL of Spiking solution 1 were placed in a 25 mL volumetric flask, and methanol was added to the mark.

Spiking solution 3:   2.5 mL of Spiking solution 1 were placed in a 25 mL volumetric flask, and methanol was added to the mark.

Four individual activated carbon tubes were spiked with separate aliquots of 10 µL Spiking solution 1 or 5 µL of Spiking solutions 1 to 3 using a syringe. Laboratory air (30–50% relative humidity) was subsequently drawn through each tube at a flow rate of 80 mL/min for two hours. This procedure covers the air concentrations given in Table 1. Six replicate determinations conducted according to the method described yielded the relative standard deviations and recoveries for TCE and PCE shown in Table 1.

Table 1.

Concentration mg/m^3		Relative standard deviation %		Recovery	
TCE	PCE	TCE	PCE	TCE	PCE
322	682	2.9	2.0	0.96	0.95
161	341	0.9	0.9	1.00	0.98
64.3	136	1.5	1.9	1.00	0.97
16.1	34.1	2.3	1.2	1.03	0.99

5.2 Limit of quantification

The absolute limit of quantification for TCE and PCE is 8 ng. This corresponds to 40 µg TCE or PCE per activated carbon tube or sample. The values were determined from the signal/noise ratio of the chromatogram.

The relative limit of quantification is
$4.2 \ mg/m^3 \triangleq 0.76 \ mL/m^3$ (ppm) for TCE and
$4.2 \ mg/m^3 \triangleq 0.60 \ mL/m^3$ (ppm) for PCE, both determined for a
9.6-litre air sample, 5 mL desorption solution and an injection volume of 1 µL.

5.3 Selectivity

The selectivity of the method depends above all on the type of column used. The column specified here has proved reliable in practice. In the presence of interfering components, it may be necessary to use a column with different separation characteristics.

6 Discussion

The loaded activated charcoal tubes can be stored in the dark for at least four weeks at room temperature without loss of adsorbed TCE or PCE.

Author: *E. Schriever*

Federation of the Employment Accidents Insurance Institutions of Germany
(Hauptverband der Berufsgenossenschaften)
Centre for Accident Prevention and Occupational Medicine
Alte Heerstraße 111, 53757 Sankt Augustin
Expert Committee Chemistry

Carcinogenic substances	Order number: BGI 505-66E
Established methods	Issue: April 1999

Method for the determination of triglycidyl isocyanurate (TGIC)

Method tested and recommended by the Berufsgenossenschaften for the determination of TGIC in working areas after discontinuous sampling.
Both personal and stationary sampling can be conducted for the assessment of working areas.

Sampling with a pump and collection on a glass-fibre filter, high-performance liquid chromatography (HPLC) after elution.
TGIC-1-HPLC
(Issue: April 1999)

IUPAC name:	1,3,5-Tris(oxiranylmethyl)-1,3,5-triazine-2,4,6(1H,3H,5H)-trione
Synonyms:	Triglycidyl isocyanurate (TGIC)
	Tris(2,3-epoxypropyl) isocyanurate
CAS No.:	2451-62-9
Molecular formula:	$C_{12}H_{15}N_3O_6$
Molar mass:	297.3 g/mol
Structural formula:	

The MAK-Collection Part III: Air Monitoring Methods, Vol. 9. DFG, Deutsche Forschungsgemeinschaft
Copyright © 2005 WILEY-VCH Verlag GmbH & Co. KGaA, Weinheim
ISBN: 3-527-31138-6

Sampling with a pump and collection on a glass-fibre filter, high-performance liquid chromatography (HPLC) after elution

Summary

This method permits the determination of TGIC concentrations in working areas averaged over the sampling time after personal or stationary sampling.

Principle:	A pump is used to draw a measured volume of air through a glass fibre filter, taking into account the definition of inhalable dust fraction [1]. The TGIC is extracted from the collected dust by treatment with tetrahydrofuran. The matrix is precipitated by adding water. The sample is then dissolved in a phosphoric acid buffer/acetonitrile mixture, filtered and analysed by liquid chromatography.

Technical data:

Limit of quantification:	absolute: 0.03 µg TGIC, equivalent to 2.5 µg per sample
	relative: 0.008 mg/m^3 for a 420-litre air sample, 2 mL sample solution and 25 µL injection volume
Selectivity:	In the presence of interfering components, the values determined may be too high. Interference can generally be eliminated by selecting different separation conditions.
Advantages:	Personal sampling and selective determinations possible
Disadvantages:	No indication of peak concentrations
Apparatus:	Pump
	Gas meter or volumetric flow meter
	Glass-fibre filter and filter holder
	Liquid chromatograph with UV detector

Detailed description of the method

Contents

1 Equipment, chemicals and solutions

1.1 Equipment

For sampling:
- Pump, suitable for flow rates of 3.5 L/min, e.g. Gilian PP5-Ex from GSM GmbH (manufacturers of measuring instruments), Gut Vellbrüggen, D-41469 Neuss, or DEHA Haan + Wittmer GmbH, D-71292 Friolzheim
- Gas meter or volumetric flow meter
- Sampling head GSP (filter holder with a sintered glass insert) e.g. from GSM GmbH or DEHA Haan + Wittmer GmbH
- Glass-fibre filters, 37 mm diameter, e.g. MN 85/90 from Macherey-Nagel

For sample preparation and analysis:
- Glass or polyethylene volumetric flask, 10 mL
- Round bottom flask, 10 mL, standard ground joint NS 14/23
- Piston pipette, 2 mL, e.g. Finnpipette Focus Fixed Volume 2000 μL from Thermo Labsystems, with the matching tips
- Adjustable volume pipettes for dosing, 50 μL–1 mL, e.g. from Hamilton

- Luer-lock syringes, 1 mL, e.g. Hamilton Series 1000
- Disposable filter, pore size 0.45 μm, e.g. Gelman Acrodisc LC13 PVDF, No. 4452
- Autosampler vials, 2 mL, e.g. from Agilent Technologies
- Rotary evaporator with a solvent-resistant vacuum pump
- Ultrapure water system, e.g. NANOpure ultrapure water system from Barnstead
- Liquid chromatograph equipped with a gradient pump, UV detector and data system, e.g. Hewlett Packard LC 1050 (with a quaternary pump, variable wavelength detector, autosampler and ChemStation software)
- pH meter
- Ultrasonic bath

1.2 Chemicals

Tetrahydrofuran (THF), analytical grade, e.g. from Merck, D-64293 Darmstadt, No. 109731
Acetonitrile, HPLC quality, e.g. from Baker, D-64347 Griesheim, No. 9017
Phosphoric acid, 85%, e.g. from Fluka, No. 79617, Sigma-Aldrich Chemie GmbH, D-82024 Taufkirchen
Potassium hydroxide pellets (KOH), analytical grade, e.g. from Merck, No. 105033
Tris(2,3-epoxypropyl) isocyanurate (TGIC), e.g. from Sigma-Aldrich, No. 379506
Ultrapure water for HPLC

1.3 Solutions

Phosphoric acid solution, H_3PO_4 concentration = 0.01 mol/L:
1,153 g phosphoric acid is placed in a 1 litre volumetric flask, and the flask is filled to the mark with ultrapure water. This solution is stable for at least 4 months at 4 °C.

Potassium hydroxide solution, KOH concentration = 15 mol/L:
8.4 g potassium hydroxide and 1.15 g sodium hydrogen carbonate are weighed into a 10 mL polyethylene volumetric flask and dissolved in ultrapure water. The flask is then filled to the mark with ultrapure water. The solution is stable for at least 6 months at room temperature.

Phosphoric acid buffer, pH 6:
The pH of the phosphoric acid buffer is adjusted to 6 with potassium hydroxide solution. This is accomplished by drop-wise addition of potassium hydroxide solution while stirring. The pH is continuously monitored with a pH meter during the process. The solution is stable for at least 2 weeks at 4 °C.

Solvent A:
Mixture of 90% (v/v) phosphoric acid buffer and 10% (v/v) acetonitrile. The solution is stable for at least one week at 4 °C.

Solvent B:
Mixture of 50% (v/v) acetonitrile and 50% (v/v) ultrapure water.

TGIC stock solutions:

Stock solution I, TGIC concentration = 400 µg/mL:
10 mg TGIC is weighed into a 25 mL volumetric flask and approx. 20 mL of Solvent A is added. The TGIC is dissolved in the ultrasonic bath and Solvent A is then added to the mark.

Stock solution II, TGIC concentration = 20 µg/mL:
1 mL Stock solution I is pipetted into a 20 mL volumetric flask and the flask is filled to the mark with the Solvent A.

Stock solution I can be stored for several weeks at –18 °C. Stock solution II must be freshly prepared from Stock solution I when needed and can then be stored for at least one week at –18 °C. It is advisable to divide Stock solution I into several portions and freeze them in separate containers immediately after preparation. The portions can then be used separately to prepare fresh batches of Stock solution II.

Stock solution II can be checked for TGIC decomposition by running a chromatogram. Decreasing signal intensity and the presence of interfering peaks indicate improper or excessive storage.

Calibration solutions:
The following calibration solutions are prepared (Table 1).

Table 1.

Solution	Stock solution II µL	Solvent A µL	TGIC µg/mL
1	100	900	2
2	250	750	5
3	500	500	10
4	750	250	15
5	1000	0	20

The specified volumes of Stock solution II and Solvent A are placed in an autosampler vial, which is subsequently sealed with a septum crimp cap.
Under the conditions described, these solutions cover a TGIC concentration range of 0.01 to 0.10 mg/m^3.
The shelf-life of the calibration solutions is very limited. They can be used for a maximum of three days if stored at 4 °C.

2 Sampling

For sampling, the GSP sampling head is fitted with a glass fibre filter and connected to the pump. The pump and filter holder are carried by a person during working hours or used in a stationary position. To take into account the definition of inhalable dust fraction [1] the flow rate is set at 3.5 L/min in accordance with the definition of inhalable dust fraction. A sampling time of two hours corresponds to an air sample volume of 420 litres.

3 Analytical determination

3.1 Sample preparation

The loaded filter is placed in a 10 mL round bottom flask and 2 mL THF is added. The flask is closed with a glass stopper and placed in the ultrasonic bath for one hour. Subsequently, 2 mL ultrapure water is added dropwise, resulting in precipitation of the matrix. The suspension is carefully evaporated to dryness in the rotary evaporator. The residue is dissolved in 2 mL Solvent A and the flask is stoppered again and placed in the ultrasonic bath for half an hour (sample solution). An aliquot of approximately 1 mL is removed with a Luer-lock syringe and filtered through a disposable filter into an autosampler vial. The vial is then sealed and its contents analysed.

3.2 Operating conditions for HPLC

The method was characterised under the following experimental conditions.

Apparatus:	Hewlett Packard LC 1050 with quaternary gradient pump, variable volume injector and autosampler, variable wavelength detector
Column:	Commercially available Hypersil column ODS C18, 5 µm, 200 × 2,1 mm, from Hewlett Packard, No. 79916 OD-552
Gradient elution:	Start: 100% Solvent A, change to 15% Solvent B over a period of 10 min, change to 100% Solvent B and hold for 5 min, change to 100% Solvent A and hold for 7 min (to condition for the next injection)
Flow rate:	0.2 mL/min
Injection volume:	25 µL
Detection wavelength:	205 nm

4 Calculations

4.1 Calibration

Aliquots of 25 µL of each of the calibration solutions exemplarily described in Section 1.3 are injected into the chromatograph and chromatograms are recorded. The peak areas determined are plotted against the corresponding concentrations of TGIC contained in the calibration solutions in order to construct the calibration curve. It is linear in the specified concentration range from 2 to 20 µg/mL.

4.2 Calculation of the analytical result

The concentration by weight of TGIC in the air sample is calculated in mg/m^3 according to Equation (1):

$$c_w = \frac{w \times 1000}{V \times \eta} \tag{1}$$

where:
c_w is the concentration by weight of TGIC in the air sample in mg/m^3
w is the weight of TGIC in mg contained in 2 mL sample solution
V is the volume of the air sample in litres
η is the recovery.

5 Reliability of the method

5.1 Accuracy and recovery

The relative standard deviation of the method was determined by spiking filters with 4 µg, 40 µg and 80 µg TGIC in replicates of 8. This was accomplished by loading the filters with the appropriate volumes of stock solution. The filters were then processed according to Section 3.1 the same day and measured the next day.
The spiking amounts of TGIC correspond to 0.01, 0.1 and 0.2 mg/m^3 for a 420-litre air sample.
The results were used for calculating relative standard deviation and recovery (Table 2).

Table 2.

Concentration mg/m^3	Relative standard deviation %	Recovery
0.01	5.7	0.75
0.1	5.0	0.73
0.2	2.5	0.76

The mean recovery is 0.75.

5.2 Limit of quantification

The limit of quantification was determined from the 10-point calibration data in accordance with DIN 32645 [2]. The maximum error of result was taken as 25%.

The absolute limit of quantification was determined as 0.03 µg TGIC, equivalent to 2.5 µg per sample. The relative limit of quantification is 0.008 mg/m^3 for a 420-litre air sample, 2 mL sample solution and an injection volume of 25 µL.

The following calibration solutions were prepared for the 10-point calibration with equal concentration steps (Table 3):

Table 3.

Solution	Stock solution II µL	Solvent A µL	TGIC µg/mL
1	100	900	2
2	200	800	4
3	300	700	6
4	400	600	8
5	500	500	10
6	600	400	12
7	700	300	14
8	800	200	16
9	900	100	18
10	1000	–	20

5.3 Selectivity

Values measured may be too high in the presence of interfering components. Should the chromatogram show overlapping peaks, then separation conditions need to be changed so that no such interference will occur. The separation conditions described here have proved reliable in practice.

5.4 Shelf-life

A suspension of 200 mg of a TGIC-containing powder coating material was prepared in approximately 20 mL *n*-heptane. Weighed glass fibre filters were each loaded with 500 µL of the solution. The filters were air-dried overnight. Differential weighing was used to determine the amount loaded. Five repeat determinations, each with triplicate filters, were performed at regular intervals over a period of 30 days. The loaded filters were stored in the laboratory at room temperature. The decrease in TGIC on the filters was markedly less than 10%.

To determine the shelf-life of pure TGIC, the substance was dissolved in Solvent A (Section 1.3) and loaded onto the filter in that form. Low recoveries were observed after only 24 hours of storage and the effect increased with the duration of storage. Therefore, it is not possible to store glass fibre filters loaded with pure TGIC.

6 Discussion

The determination of pure TGIC does not require treatment of the sample with THF (Section 3.1) during sample preparation. When determining TGIC in TGIC-containing powder coating materials, this is a necessary step in sample preparation, since comparative studies have demonstrated that recovery losses of up to 70% are possible if THF treatment is omitted.

7 References

[1] *Deutsches Institut für Normung e.V.* (DIN) (1993): DIN EN 481 – Arbeitsplatzatmosphäre – Festlegung der Teilchengrößenverteilung zur Messung luftgetragener Partikel. Beuth-Verlag, Berlin.
[2] *Deutsches Institut für Normung e.V.* (DIN) (1994): DIN 32645 – Nachweis-, Erfassungs- und Bestimmungsgrenze. Beuth Verlag, Berlin.

Author: *J.-U. Hahn*

Zirconium

Method number   1

Application   Air analysis

Analytical principle Energy dispersive X-ray fluorescence

Completed in   May 2003

Summary

The present method uses energy dispersive X-ray fluorescence to analyse dusts containing zirconium compounds collected on membrane filters, quantifying them against calibration filters. The method offers a rapid and reliable means of determining the concentration of zirconium in workplace air.

Characteristics of the method

Accuracy (determined from the counting statistics for the calibration samples):

Residual standard deviation:	$s_y = 31.9$ counts/s
Standard deviation of the method:	$s_{x0} = 1.30$ µg/cm^2
Relative standard deviation of the method:	$V_{x0} = 4.539$ µg/cm^2
Recovery:	$\eta = 0.95-1.05$ (95–105%)
Limit of quantification:	0.096 mg/m^3 for a sample volume of 120 litres
Sampling recommendation:	Sampling time: 2 hours Air sample volume: 120 litres

The MAK-Collection Part III: Air Monitoring Methods, Vol. 9. DFG, Deutsche Forschungsgemeinschaft
Copyright © 2005 WILEY-VCH Verlag GmbH & Co. KGaA, Weinheim
ISBN: 3-527-31138-6

Zirconium [CAS No. 7440-67-7]

Zirconium has a molar mass of 91.22 g/mol and a melting point of 1852 °C. Its density is 6.5 g/cm^3. Zirconium forms dense, adherent protective coatings with many aggressive acidic or basic media, the layers consisting of stable, high-melting oxides. Therefore, zirconium is very resistant to corrosion. Zirconium dioxide (ZrO_2) is primarily used in highly refractory ceramics for the manufacture of bricks and fittings. Zirconium silicate, or zircon ($ZrSiO_4$), is used mainly in the manufacture of fine ceramic masses and ceramic insulation materials. Very finely ground zircon sand (grain size 3–5 μm) is used as a polishing agent (e. g. for glass polishing) [1].

Zirconium and insoluble zirconium compounds have been assigned a MAK value of 1 mg/m^3 for inhalable dust, whereas soluble zirconium compounds cannot, at present, be assigned a MAK value [2]. There is a risk of sensitisation of the skin and airways by zirconium and its compounds. The TRGS 900 gives threshold limit values of 1.0 and 2.5 mg/m^3 [3].

Authors: *M. von Döhren, M. Tschickardt*
Examiners: *J. U. Hahn, C. Hagemann*

Zirconium

Method number   1

Application   Air analysis

Analytical principle Energy dispersive X-ray fluorescence

Completed in   May 2003

Contents

1 General principles

The present method uses energy dispersive X-ray fluorescence to analyse dusts containing zirconium compounds collected on membrane filters, quantifying them against calibration filters. The method offers a rapid and reliable means of determining the concentration of zirconium in workplace air.

A pump is used to draw a measured volume of air through a pre-weighed membrane filter. The zirconium and zirconium compounds present in the collected dust are determined by energy dispersive X-ray fluorescence using the target technique.

A suitably sized filter segment is punched out of the membrane filter and placed on the sample holder of the analyser with the unloaded surface facing the X-ray tube.

Quantification is achieved using calibration filters prepared in house by loading them dropwise with zirconium-containing solutions. Care must be taken to ensure best possible homogeneous spreading. The weight per area of filter ($\mu g/cm^2$) is determined from the count rates of the zirconium lines by comparison with the calibration filters. The concentration by weight present in the air is then calculated, taking into account the whole area of the loaded filter and the volume of air drawn through the filter.

2 Equipment, chemicals and solutions

2.1 Equipment

Energy dispersive X-ray spectrometer (e.g. Kevex Analyst 770, from GETAC Instrumentenbau GmbH, Mainz)

Dispenser with a 1000 μL syringe and hand probe button (e.g.. Microlab M from Hamilton Deutschland GmbH, Martinsried)

Desiccator

Membrane filter of the Millipore AAWG type, diameter 37 mm, gridded, pore size 0.8 μm

31 mm Punch for cutting segments out filters, zirconium-free

Sampling pump, flow rate 1 L/min (e.g. Alpha-1, from Du Pont, USA) GSP sampling system with inlet cone 1.0 according to BIA (e.g. from GSM Schadstoffmesstechnik, Neuss)

Laboratory balance

10, 100 and 1000 mL Volumetric flasks

2.2 Chemicals

Zirconium standard for atomic absorption spectroscopy, ready-for-use ampoule ($c(Zr) = 1.000$ g/L prepared from $ZrOCl_2$, HCl and H_2O), from Fluka

Distilled water

2.3 Solutions

Stock solution 1: The contents of the ready-for-use ampoule are rinsed into a 1 litre volumetric flask, which is then filled to the mark with water.

Stock solution 2: The contents of the ready-for-use ampoule are rinsed into a 100 mL volumetric flask, which is then filled to the mark with water.

2.4 Calibration standards

Preparation of the calibration solutions:
Aliquots of stock solutions 1 and 2 are taken according to the following pipetting schemes and each filled to 10 mL with distilled water (Tables 1 and 2). The calibration solutions are used for the preparation of the calibration filters.

Procedure:
The filter is placed on a 30-mm neck Erlenmeyer flask. The syringe of the Hamilton dispenser is filled with the appropriate calibration solution, which is dispensed dropwise onto the filter surface such that each square is loaded with 1 µL of the appropriate calibration solution. On completion of dropwise loading, the filter is allowed to dry overnight.

Table 1. Pipetting scheme for calibration from stock solution 1.

Calibration solution No.	Volume of stock solution 1 mL	Final volume of the calibration solution mL	Concentration g/L	Loading µg/cm^2
1	0.25	10	0.025	0.26
2	0.50	10	0.050	0.52
3	1	10	0.100	1.04
4	2	10	0.200	2.08
5	4	10	0.400	4.16
6	6	10	0.600	6.24
7	8	10	0.800	8.32
8	10	10	1.000	10.4

Table 2. Pipetting scheme for calibration from stock solution 2.

Calibration solution No.	Volume of stock solution 2 mL	Final volume of the calibration solution mL	Concentration g/L	Loading µg/cm^2
9	2	10	2.000	20.81
10	4	10	4.000	41.62
11	6	10	6.000	62.43
12	8	10	8.000	83.25
13	10	10	10.000	104.06

The loading on the calibration filter is calculated according to Equation (1):

$$L = \frac{C \times V_{cs}}{F} \qquad (1)$$

where:
L is the loading on the filter in $\mu g/cm^2$
V_{cs} is the volume of the calibration solution per filter square in L
C is the concentration of the calibration solution in g/L
F is the area of each filter square (here: 0.0961 cm^2)

3 Sampling and sample preparation

Pumps used for sampling are of the "Alpha-1" type from Du Pont (supplied in Germany by DEHA Haan + Wittmer GmbH, D-71292 Friolzheim). Sampling is conducted using GSP sampling head 1.0 for inhalable dust at a flow rate of 1 L/min. The duration of sampling is between 30 minutes and two hours. Dust is collected on pre-weighed AAWG-type membrane filters from Millipore with a pore size of 0.8 μm. Sampling can be conducted as personal or stationary sampling.

After sampling, the filters are conditioned in a desiccator and weighed again to determine dust loading. A 31-mm diameter segment is cut out of the dust-loaded membrane filter using a zirconium-free punch.

These are then placed on the sample holder of the analyser with the unloaded surface facing the X-ray tube.

4 Instrumental operating conditions

Apparatus:	Kevex Analyst 770, from GETAC Instrumentenbau GmbH, D-55126 Mainz
X-ray tube:	Rhodium anode side window X-ray tube, 200 W power output without filter, indirect excitation with an Ag target
X-ray generator:	45 kV; 2.5 mA
Counting time:	200 s
Detector:	Si(Li) semi-conductor detector
	10 eV/channel
	2048 channels
	Multichannel analyser 20 keV

5 Analytical determination

The filters are analysed and the spectra stored electronically (Figure 1). The spectra of the sample filters and calibration filters are used to calculate the intensities of the spectral lines which are characteristic of zirconium.

6 Calibration

The calibration curve is constructed by plotting the intensities of the zirconium signals determined by the XRF data system against the corresponding loadings on the calibration filters (Figure 2). The calibration curve is linear up to about 100 $\mu g/cm^2$ and then changes into a second-order function.
In order to check calibration, a calibration filter in the range of the MAK value is included in every analytical run (Table 3).

Table 3. Signal intensities of the calibration filters, mean values of $n = 6$ determinations.

Loading $\mu g/cm^2$	Zirconium peak counts/s	Standard deviation (s)	Relative standard deviation (s_{rel})
0	0.529	0.27	50.7
0.26	6.719	0.43	6.4
0.52	14.082	1.28	9.1
1.04	26.912	1.66	6.2
2.08	54.771	2.48	4.5
4.16	113.070	5.76	5.1
6.24	168.540	3.55	2.1
8.32	227.380	7.10	3.1
10.40	269.580	16.23	6.0
20.81	445.36	19.93	4.50
41.62	973.91	73.09	7.50
62.43	1518.45	106.10	7.00
83.25	2052.96	27.90	1.40
104.06	2571.32	61.68	2.40

7 Calculation of the analytical result

Using the intensities obtained, the corresponding weight X in $\mu g/cm^2$ is read from the calibration curve. The corresponding concentration by weight (ρ) is calculated according to the following equation:

$$\rho = \frac{X \times F}{V} \tag{2}$$

At 20 °C and 1013 hPa:

$$\rho_0 = \rho \times \frac{273 + t_a}{293} \times \frac{1013}{p_a} \tag{3}$$

where:

ρ is the concentration of a component in mg/m^3

ρ_0 is the concentration in mg/mg^3 at 20 °C and 1013 hPa

X is the weight of the component on the filter in $\mu g/cm^2$

F is the loaded filter area, here: $8.55\ cm^2$

t_a is the temperature during sampling in °C

p_a is the atmospheric pressure during sampling in hPa

V is the air sample volume in litres (calculated from the flow rate and the sampling time)

8 Reliability of the method

8.1 Accuracy

8.1.1 Precision

The counting statistics for the drip-loaded calibration samples were used to calculate the precision data. The following data were obtained:

Residual standard deviation:	$s_y = 31.9$ counts/s
Standard deviation of the method:	$s_{x0} = 1.30\ \mu g/cm^2$
Relative standard deviation of the method:	$V_{x0} = 4.539\ \mu g/cm^2$

Furthermore, the filters drip-loaded for calibration purposes were stored under ambient conditions. After 6 months, 3 parallel filters were re-analysed for each of 3 concentrations [4] (10.40, 62.43 and 104.04 $\mu g/cm^2$). The data shown in Table 4 were obtained.

Table 4. Relative standard deviation s_{rel}, $n = 3$ determinations.

Filter loading $\mu g/cm^2$	Mean value $\mu g/cm^2$	Relative standard deviation (s_{rel}) %	Recovery %
10.40	11.31	1.6	109
62.43	64.29	1.5	103
104.04	106.17	2.5	102

8.1.2 Accuracy of the mean

In order to determine the accuracy of the mean, actual samples loaded with zirconium silicate were analysed gravimetrically and by total reflection X-ray fluorescence (TRXF). Gravimetric recovery was determined to be 95 to 105%.
For analysis by TRXF, the filters were dissolved in acetone, internal standards were added and an aliquot was applied to a sample carrier and dried. Six filters were analysed for each of three different dust loadings (Table 5).

Table 5. Results from comparative EDXRF/TRXF studies.

Sample	Laboratory A (EDXRF)		Laboratory B (TRXF)	
	Mean value mg/m^3	Standard deviation (s)	Mean value mg/m^3	Standard deviation (s)
1	10.18	0.947	8.25	0.889
2	0.21	0.017	0.21	0.020
3	0.03	0.001	0.04	0.012

8.2 Limit of quantification

Determination of the limit of quantification according to DIN 32645 ("blank value method") yielded 0.096 mg/m^3 for the analysis of 15 blank filters and an air sample volume of 120 litres [5].

8.3 Shelf-life

Shelf-life tests play a minor role in the case of zirconium. As a rule there are no storage losses.

8.4 Interference

Depending on dust weights collected, absorption effects may occur at high weights. The analytical result is affected by inhomogeneous distribution of the collected particles. In such a situation, a more representative result is obtained by repeating analysis of the filters and turning them by 90° for each measurement.
Dusts of unknown composition may be associated with spectral interferences, which can hamper exact quantification.
When using membrane filters, as described in the present method, the capacity of the sample carrier is limited to approx. 225 µg dust/cm^2. At higher loadings, there is detachment of the collected dust particles. Loading was verified gravimetrically.

9 Discussion of the method

The method enables the determination of zirconium compounds up to a loading of approx. 100 $\mu g/cm^2$, corresponding to a concentration of 7.4 mg/m^3.
Calibration filters loaded with 208 or 312 $\mu g/cm^2$ yielded a second-order regression calibration curve. Shortening the sampling time provides a solution to this problem.

10 References

[1] *Ullmanns Encyklopädie der technischen Chemie* (1972) Zirkonium, Band 24: 681–702. VCH-Verlag, Weinheim

[2] *Deutsche Forschungsgemeinschaft* (2004) List of MAK und BAT Values 2004. Commission for the Investigation of Health Hazards of Chemical Compounds in the Work Area. Report No. 40. Wiley-VCH Verlag, Weinheim

[3] *Bundesministerium für Wirtschaft und Arbeit* (2002) TRGS 900 „Grenzwerte in der Luft am Arbeitsplatz-Luftgrenzwerte", Ausgabe Oktober 2000, zuletzt geändert BarbBl. 03/2003

[4] *Europäisches Komitee für Normung* (CEN) (1994) DIN EN 482 – Arbeitsplatzatmosphäre – Allgemeine Anforderungen an Verfahren zur Messung von chemischen Arbeitsstoffen. Brüssel 1994. Beuth Verlag, Berlin

[5] *Deutsches Institut für Normung e.V.* (DIN) (1994) DIN 32645 – Chemische Analytik-Nachweis-, Erfassungs- und Bestimmungsgrenze. Beuth Verlag, Berlin

Authors: *M. von Döhren, M. Tschickardt*
Examiners: *J. U. Hahn, C. Hagemann*

Fig. 1. Energy dispersive X-ray fluorescence (EDXRF) spectrum of a filter (actual sample).

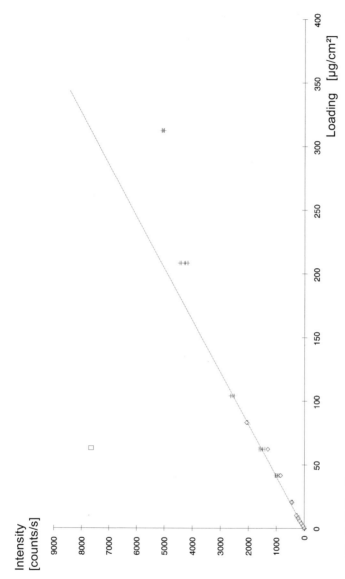

Fig. 2. Calibration curve.

Members, Guests and ad hoc Experts

of the Working Subgroup "Analyses of Hazardous Substances in Air of Work Area" of the Commission for the Investigation of Health Hazards of Chemical Compounds in the Work Area of the Deutsche Forschungsgemeinschaft (Status November 2004)

Leader:

Prof. Dr. Dr. H. Parlar
Technische Universität München
Wissenschaftszentrum Weihenstephan für Ernährung, Landnutzung und Umwelt
Lehrstuhl für Chemisch-Technische Analyse und Chemische Lebensmitteltechnologie
Weihenstephaner Steig 23
85350 Freising-Weihenstephan

Members and Permanent Guests:

Prof. Dr. J. Angerer
Institut für Arbeits-, Sozial- und Umweltmedizin
der Universität Erlangen-Nürnberg
Schillerstraße 25/29
91054 Erlangen

Prof. Dr. E. Hallier
Georg-August-Universität Göttingen
Zentrum Umwelt- und Arbeitsmedizin
Abt. für Arbeits- und Sozialmedizin
Waldweg 37
37073 Göttingen

Dr. R. Hebisch
Bundesanstalt für Arbeitsschutz und Arbeitsmedizin
Gruppe 4.1 "Belastungen am Arbeitsplatz"
Friedrich-Henkel-Weg 1–25
44149 Dortmund

Guests:

Dr. M. Ball
Ergo-Forschungsgesellschaft GmbH
Geierstraße 1
22305 Hamburg

The MAK-Collection Part III: Air Monitoring Methods, Vol. 9. DFG, Deutsche Forschungsgemeinschaft
Copyright © 2005 WILEY-VCH Verlag GmbH & Co. KGaA, Weinheim
ISBN: 3-527-31138-6

Dr. D. Breuer
Berufsgenossenschaftliches Institut für Arbeitssicherheit-BIA
Alte Heerstr. 111
53754 Sankt Augustin

Dr. C. Habarta
Bayerisches Landesamt für Arbeitsmedizin u. Sicherheitstechnik
Pfarrstraße 3
80538 München

Dr. J. U. Hahn
Berufsgenossenschaftliches Institut für Arbeitssicherheit-BIA
Alte Heerstr. 111
53754 St. Augustin

Dr. E. Hellpointner
Bayer CropScience AG
RD-Development/MEF, Building 6660
Alfred-Nobel-Str. 50
40789 Monheim am Rhein

Dr. W. Kleiböhmer
Institut für Chemo- und Biosensorik e.V.
Mendelstr. 7
48149 Münster

Prof. Dr. A. Kettrup
Lehrstuhl für Ökologische Chemie und Umweltanalytik
Department für Biowissenschaftliche Grundlagen
Weihenstephaner Steig 23
85350 Freising- Weihenstephan

Dr. W. Krämer
BASF-AG
Labor für Umweltanalytik
Abt. DUU/OU – Z 570
67056 Ludwigshafen

Dr. N. Lichtenstein
Berufsgenossenschaftliches Institut für Arbeitssicherheit-BIA
Alte Heerstr. 111
53754 St. Augustin

Dr. C.-P. Maschmeier
Landesamt für Arbeitsschutz des Landes Sachsen-Anhalt
Kühmauer Str. 70
06846 Dessau

Dr. R. Meyer zu Reckendorf
Zentrallabor
SGL Carbon GmbH
Werner-von-Siemens Str. 18
86405 Meitingen

Dipl.-Ing. K.H. Pannwitz
Dräger Safety AG & Co. KGaA
Abteilung Analysentechnik
Revalstraße 1
23560 Lübeck

Dipl.-Ing. M. Tschickardt
Landesamt für Umwelt, Wasserwirtschaft und Gewerbeaufsicht
Kaiser-Friedrich-Str. 7
55116 Mainz

Prof. Dr. W. Riepe
Universität Salzburg Inst. Chemie u. Biochemie
Hellbrunnerstr. 34
5020 Salzburg
AUSTRIA

Ad hoc Experts:

Dr. D. Franke
AQura GmbH, Analytik-Center
Gebäude/PB: 9015/10
Paul-Baumannstr. 1
45764 Marl

Dr. H. Fricke
Bergbau-Berufsgenossenschaft
Hauptverwaltung Hunscheidtstraße 18
44789 Bochum

Dr. D. Kotzias
European Commission-Joint Research Centre
Institute for Health and Consumer Protection
Unit "Physical and Chemical Exposure"
21020 Ispra (VA)
ITALY

Dipl.-Ing. U. Lehnert
Wehrwissenschaftliches Institut für Werk, Explosiv und Betriebsstoffe (WIWEB)
Landshuterstr. 70
85435 Erding

Dr. U. Lewin-Kretzschmar
BG Chemie
Berufsgenossenschaft der chemischen Industrie
Analytisches Labor Leuna
Rudolf-Breitscheid-Str. 18 E
06237 Leuna

Dr. B. Schneider
Clariant GmbH – Division Feinchemikalien
Analytik Intermedites
Stroofstr. 27
65933 Frankfurt/M

Scientific secretariat:

Dr. M.R. Lahaniatis
Technische Universität München
Wissenschaftszentrum Weihenstephan für Ernährung, Landnutzung und Umwelt
Lehrstuhl für Chemisch-Technische Analyse u. Chemische Lebensmitteltechnologie
Weihenstephaner Steig 23
85350 Freising-Weihenstephan

Dr. R. Schwabe
Kommissions-Sekretariat
Hochenbacherstr. 15–17
85350 Freising-Weihenstephan

Members of the Analytical Working Group
of the Board of Experts "Chemistry"

Dr. Brock   Berufsgenossenschaft der chemischen Industrie
                           Heidelberg (Chairman)

Dr. Feige   Merck KGaA
                           Darmstadt

Dr. Keller   Consultant to the Berufsgenossenschaft
                           der chemischen Industrie
                           Leverkusen

Dr. Krämer   BASF AG
                           Ludwigshafen

Dr. Lauterwald   Landesamt für Umweltschutz und Gewerbeaufsicht
                           Mainz

Dr. Lewin-Kretzschmar   Berufsgenossenschaft der chemischen Industrie
                           Analytisches Labor
                           Leuna

Dr. Lichtenstein   Berufsgenossenschaftliches Institut
                           für Arbeitssicherheit – BIA
                           Sankt Augustin

Dr. Schriever   Volkswagen AG
                           Wolfsburg

Dipl.-Ing. Sporenberg   Bayer AG
                           Leverkusen

Dr. Wildenauer   Degussa AG
                           Trostberg

Contents of Volumes 1–9

The MAK-Collection Part III: Air Monitoring Methods, Vol. 9. DFG, Deutsche Forschungsgemeinschaft
Copyright © 2005 WILEY-VCH Verlag GmbH & Co. KGaA, Weinheim
ISBN: 3-527-31138-6

CAS No.	Substance	Method	
1118-46-3	Butyltin trichloride	Organotin compounds	
1305-62-0	Calcium hydroxide	Alkali metal hydroxides and alkaline earth hydroxides	
1309-64-4	Antimony trioxide	Antimony trioxide	7
1310-58-3	Potassium hydroxide	Alkali metal hydroxides and alkaline earth hydroxides	8
1310-65-2	Lithium hydroxide	Alkali metal hydroxides and alkaline earth hydroxides	8
1310-73-2	Sodium hydroxide	Alkali metal hydroxides and alkaline earth hydroxides	8
1461-22-9	Tributyltin chloride	Organotin compounds	3
1461-25-2	Tetrabutyltin	Organotin compounds	3
1746-81-2	3-(4-chlorophenyl)-1-methoxy-1-methylurea (Monolinuron)	Urea herbicides	3
1912-24-9	Atrazine	Atrazine	8
2243-62-1	1,5-Diaminonaphthalene	1,5-Diaminonaphthalene	5
2451-62-9	Triglycidyl isocyanurate	Triglycidyl isocyanurate (TGIC)	9
3060-89-7	3-(4-bromophenyl)-1-methoxy-1-methylurea (Metobromuron)	Urea herbicides	3
4170-30-3	trans-2-Butenal	2-Butenal	9
4184-79-6	5,6-Dimethylbenzotriazole	Benzotriazoles	8
7439-92-1	Lead	Lead	1
7440-02-0	Nickel	Nickel	1, 7
7440-31-5	Tin	Total tin	2
		Organotin compounds	3
7440-43-9	Cadmium	Cadmium	4
7440-47-3	Chromium	Chromium	1
7440-48-4	Cobalt	Cobalt	1
7440-67-7	Zirconium	Zirconium	9
7446-09-5	Sulfur dioxide	Sulfur dioxide	8
7647-01-0	Hydrochloric acid	Volatile inorganic acids	6
7664-38-2	Phosphoric acid	Inorganic acid mists	6
7664-39-3	Hydrofluoric acid	Hydrogen fluoride and fluorides	9
7664-41-7	Ammonia	Ammonia	2, 9
7664-93-9	Sulfuric acid	Inorganic acid mists	6
		Sulfuric acid	9
7697-37-2	Nitric acid	Volatile inorganic acids	6
7722-84-1	Hydrogen peroxide	Hydrogen peroxide	8
7803-12-2	Phosphine	Phosphine	5
10024-97-2	Dinitrogen oxide	Dinitrogen oxide	2
10028-15-6	Ozone	Ozone	3
10035-10-6	Hydrobromic acid	Volatile inorganic acids	6